SCIENCE
and the Akashic Field

"Ervin Laszlo presents readers with a tour de force, nothing less than a theory of everything. This book introduces such provocative concepts as the 'A-field' and the 'informed universe,' making the case that a complete understanding of reality is woefully lacking without them. Readers of this book will never view the universe in quite the same way again."

STANLEY KRIPPNER, PH.D.,
PROFESSOR OF PSYCHOLOGY, SAYBROOK GRADUATE SCHOOL,
AND AUTHOR AND COEDITOR OF *VARIETIES OF ANOMALOUS EXPERIENCE*

"Over the last 30 years, Ervin Laszlo has consistently been at the forefront of scientific inquiry, exploring the frontiers of knowledge with insight, wisdom, and integrity. With *Science and the Akashic Field* he takes another quantum leap forward in our understanding of the universe and ourselves. This enthralling vision of mind, science, and universe is essential reading for the 21st century."

ALFONSO MONTUORI, PH.D.,
CALIFORNIA INSTITUTE OF INTEGRAL STUDIES,
AND AUTHOR OF *CREATORS ON CREATING*

"It is rare indeed that a revolution in thought can open our eyes to a new universe that transforms our inner experience as well as our relationships with others and even with the cosmos. Martin Buber did it with *I and Thou*. Now, Ervin Laszlo, one of the most profound minds of our generation, has given us a great gift in this readable book that explores how we are connected to each other in fields of resonance that penetrate to the deepest levels of being."

ALLAN COMBS, PH.D.,
PROFESSOR OF PSYCHOLOGY, UNIVERSITY OF NORTH CAROLINA
AT ASHEVILLE, AND AUTHOR OF *THE RADIANCE OF BEING*

"If you ever wanted to hold the universe in your hand, pick up this book. You can hardly do better than join cosmologist Ervin Laszlo in the ultimate quest: for a theory of everything."

CHRISTIAN DE QUINCEY, PH.D.,
PROFESSOR OF PHILOSOPHY, JOHN F. KENNEDY UNIVERSITY,
EDITOR OF INSTITUTE OF NOETIC SCIENCES' *IONS REVIEW*,
AND AUTHOR OF *RADICAL NATURE: REDISCOVERING THE SOUL OF MATTER*

"In this impressive and transformative work Laszlo brings the reader into an integral worldview for our time. The reader who encounters this book will be irrevocably transformed and will henceforth experience the world through a global lens."

ASHOK GANGADEAN, PH.D.,
PROFESSOR OF PHILOSOPHY, HAVERFORD COLLEGE,
FOUNDER-DIRECTOR OF THE GLOBAL DIALOGUE INSTITUTE,
AND AUTHOR OF *THE AWAKENING OF THE GLOBAL MIND*

"In a visionary way based on profound knowledge of modern science, Laszlo creates a genuine architecture of human and cosmic evolution. He provides the bridge between all the different puzzle-stones of science and unifies them in a most remarkable and bold 'integral theory of everything.'"

FRITZ-ALBERT POPP, PH.D.,
DIRECTOR OF THE INTERNATIONAL INSTITUTE OF BIOPHYSICS,
AND EDITOR OF *RECENT ADVANCES IN BIOPHOTON RESEARCH*

"This is one of the most important books to be published in the last decades. Ervin Laszlo's *Science and the Akashic Field* has the power and coherence to explain the major phenomena of cosmos, life, and mind as they occur at the various levels of nature and society. In demonstrating that an information field is a fundamental factor in the universe, Ervin Laszlo catalyzes a radical paradigm-shift in the contemporary sciences."

IGNAZIO MASULLI, PH.D.,
PROFESSOR OF HISTORY, UNIVERSITY OF BOLOGNA, ITALY,
AND COAUTHOR OF *THE EVOLUTION OF COGNITIVE MAPS*

"Laszlo's book opens the way toward a great synthesis. Whoever reads Laszlo's book witnesses the greatest awakening of the human spirit. Not since Plato and Democritus has there been such a transformation in the history of thought!"

László Gazdag, Ph.D.,
physicist and professor of Social Sciences, Science University of
Pécs, Hungary, and author of *Beyond the Theory of Relativity*

"In his admirable 40-year quest for an integral theory of everything, Laszlo has not restricted himself to physics but presented a coherent global hypothesis of connectivity between quantum, cosmos, life and consciousness. I cannot think of anyone else who is better prepared and more able, than the genuine post-modern Renaissance Man Laszlo, to offer a vision that is imaginative, but not imaginary, a vision where all things are connected with all other things and nothing disappears without a trace."

Zev Naveh, Ph.D.,
professor emeritus, Israel Institute of Technology,
and author of *Landscape Ecology*

"Is everything that ever happened on this earth recorded in some huge, ultra-dimensional information bank? Are some of us occasionally able to tap into it with some facility, and perhaps all of us to some extent now and then during our lives? *Science and the Akashic Field* provides the pioneering scientific answer to these and many other fundamental questions our species faces at this critical time in human evolution."

David Loye, Ph.D.,
former research director of the Program on Psychosocial
Adaptation and the Future, UCLA School of Medicine,
and author of *An Arrow Through Chaos*

"*Science and the Akashic Field* shows clearly that science is poised at the threshold of a new paradigm. The new vision offers humanity the perspective of more peace and security, not as an idealistic goal but as a reflection of reality."

Jurriaan Kamp,
editor in chief of *Ode Magazine*,
and author of *Because People Matter*

SCIENCE

and the
Akashic Field

An Integral Theory of Everything

ERVIN LASZLO

Second Edition

Inner Traditions
Rochester, Vermont

Inner Traditions
One Park Street
Rochester, Vermont 05767
www.InnerTraditions.com

Library of Congress Cataloging-in-Publication Data
Laszlo, Ervin, 1932–
  Science and the akashic field : an integral theory of everything / Ervin Laszlo. — 2nd ed.
     p. cm.
  Includes bibliographical references and index.
  ISBN-13: 978-1-59477-181-1 (pbk.)
  ISBN-10: 1-59477-181-2 (pbk.)
  1. Akashic records. 2. Parapsychology and science. I. Title.
  BF1045.A44L39 2007
  501—dc22
                                            2007000623

Printed and bound in the United States by Lake Book Manufacturing

10 9 8 7 6 5 4 3

Text design and layout by Rachel Goldenberg
This book was typeset in Sabon, with Avenir as a display typeface

To send correspondence to the author of this book, mail a first-class letter to the author c/o Inner Traditions • Bear & Company, One Park Street, Rochester, VT 05767, and we will forward the communication.

*For Christopher and Alexander, who
continue to comprehend, connect,
and co-create—with love*

⚛

Contents

PART TWO

THE IN-FORMED UNIVERSE

*Perennial Questions and Fresh Answers from the
Integral Theory of Everything*

Akasha ($\bar{a}$ • k$\bar{a}$ • sha) is a Sanskrit word meaning "ether": all-pervasive space. Originally signifying "radiation" or "brilliance," in Indian philosophy akasha was considered the first and most fundamental of the five elements—the others being vata (air), agni (fire), ap (water), and prithivi (earth). Akasha embraces the properties of all five elements: it is the womb from which everything we perceive with our senses has emerged and into which everything will ultimately redescend. "The Akashic Record" (also called "The Akashic Chronicle") is the enduring record of all that happens, and has ever happened, in the whole of the universe.

Introduction

A Meaningful Scientific
Worldview for Our Time

Notwithstanding a widespread view, science is not only a collection of observations, measurements, and mathematical formulas; it is also a source of insight into the way things are in the world. Great scientists are concerned not only with the *how* of the world—the way things *work*—but also what the things of this world *are*, and *why* they are the way we find them.

It is indisputable, of course, that in the mainstream of the science community researchers are often more concerned with making their equations pan out than with the meaning they can attach to them. But this is not so in the case of leading theoreticians. The cosmological physicist Stephen Hawking, for example, is keenly interested in making clear the meaning of his theories, even though this is by no means an easy task and he does not always succeed. Shortly after the publication of his *A Brief History of Time*, a feature story appeared in the *New York Times* entitled, "Yes Professor Hawking, but what does it mean?" The question was much to the point: Hawking's theory of time and the universe is complex, and its meaning is by no means transparent. Yet his attempts to make it so are noteworthy, and worthy of being emulated.

Evidently, the search for a meaningful view of the world is not confined to science. It is entirely fundamental for the human mind; it is as old as civilization. For as long as people looked at the Sun, the Moon, and the starry sky above, and at the seas, the rivers, the hills, and the forests below, they wondered where it all came from, where it all is going, and what it all means. In the modern world, great scientists wonder as well. Some have a deep mystical streak; Newton and Einstein are prime examples. As the Canadian physicist David Peat affirmed, the leading researchers accept the challenge of finding meaning in and through science.

"Each of us is faced with a mystery," Peat began his book *Synchronicity.* "We are born into this universe, we grow up, work, play, fall in love, and at the ends of our lives, face death. Yet in the midst of all this activity we are constantly confronted by a series of overwhelming questions: What is the nature of the universe and what is our position in it? What does the universe mean? What is its purpose? Who are we and what is the meaning of our lives?" Science, Peat claims, attempts to answer these questions, since it has always been the province of the scientist to discover how the universe is constituted, how matter was first created, and how life began.

There are many scientists who reflect on these questions, but some of them come to different conclusions. The physicist Steven Weinberg is adamant that the universe as a physical process is meaningless; the laws of physics offer no discernible purpose for human beings. "I believe there is no point that can be discovered by the methods of science," he said in an interview. "I believe that what we have found so far—an impersonal universe which is not particularly directed towards human beings—is what we are going to continue to find. And that when we find the ultimate laws of nature they will have a chilling, cold, impersonal quality about them."

This split in the leading scientists' view of the world has deep cultural roots. It reflects what the historian of civilization Richard Tarnas called Western civilization's "two faces." One face is that of progress, the other, of fall. The more familiar face is the account of a long and heroic journey from a primitive world of dark ignorance, suffering, and limitation to the bright modern world of ever-increasing knowledge, freedom, and well-being, made possible by the sustained development of human reason and, above all, of scientific knowledge and technological skill. The other face is the story of humanity's fall and separation from the original state of oneness with nature and cosmos. While in the primordial condition humans possessed an instinctive knowledge of the sacred unity and profound interconnectedness of the world, a deep schism arose between humankind and the rest of reality with the ascendance of the rational mind. The nadir of this development is reflected in the current ecological disaster, moral disorientation, and spiritual emptiness.

Contemporary civilization displays both the positive and the negative faces. Some, like Weinberg, express the negative face of Western civilization. For them, meaning resides in the human mind alone: the world itself is impersonal, without purpose or intention. Others, like Peat, insist that though the universe has been disenchanted by science, it is reenchanted again in light of the latest findings.

The latter view is gaining ground. At its cutting edge, the new cosmology discovers a world where the universe does not end in ruin, and the new physics, the new biology, and the new consciousness research recognize that life and mind are integral elements in the world, and not accidental by-products.

In this book I discuss the origins and the essential elements of the worldview now emerging at the cutting edge of the new sciences. I explore why and how it is surfacing in physics and in cosmology, in the biological sciences, and in the new field of consciousness research. Then I highlight the crucial feature of the emerging worldview: the revolutionary discovery that at the roots of reality there is not just matter and energy, but also a more subtle but equally fundamental factor, one that we can best describe as active and effective information: "in-formation."

In-formation, I claim, links all things in the universe, atoms as well as galaxies, organisms the same as minds. This discovery transforms the fragmented world-concept of the mainstream sciences into an integral, holistic worldview. It opens the way toward the elaboration of a theory that has been much discussed but until recently has not been truly achieved: an integral theory not just of one kind of things, but of all kinds—an *integral theory of everything*.

An integral theory of everything would bring us closer to understanding the real nature of all the things that exist and evolve in space and time, whether they are atoms or galaxies or mice and men. It gives us an encompassing and yet scientific view of ourselves and of the world; a view that we very much need in these times of accelerating change and mounting disorientation.

PART ONE

THE FOUNDATIONS OF AN INTEGRAL THEORY OF EVERYTHING

How Information Connects Everything to Everything Else

Come,
sail with me on a quiet pond.
The shores are shrouded,
the surface smooth.
We are vessels on the pond
and we are one with the pond.

A fine wake spreads out behind us,
traveling throughout the misty waters.
Its subtle waves register our passage.

Your wake and mine coalesce,
they form a pattern that mirrors
your movement as well as mine.
As other vessels, who are also us,
sail the pond that is us as well,
their waves intersect with both of ours.
The pond's surface comes alive
with wave upon wave, ripple upon ripple.
They are the memory of our movement;
the traces of our being.

The waters whisper from you to me and from me to you,
and from both of us to all the others who sail the pond:

Our separateness is an illusion;
we are interconnected parts of the whole—
we are a pond with movement and memory.
Our reality is larger than you and me,
and all the vessels that sail the waters,
and all the waters on which they sail.

ONE

The Challenge of an Integral Theory of Everything

In this opening chapter we discuss the challenge of creating a "TOE"—a theory of everything. A theory that deserves this name must be a theory truly of *everything*—an integral theory of all the kinds of things we observe, experience, and encounter, whether they are physical things, living things, social and ecological things, or "things" of mind and consciousness. Such an "I-TOE" can be achieved—as this chapter and those that follow will show.

There are many ways of comprehending the world: through personal insight, mystical intuition, art, and poetry, as well as the belief systems of the world's religions. Of the many ways available to us, there is one that is particularly deserving of attention, for it is based on repeatable experience, follows a rigorous method, and is subject to ongoing criticism and assessment. It is the way of science.

Science, as a popular newspaper column tells us, matters. It matters not only because it is a source of the new technologies that are shaping our lives and everything around us, but also because it suggests a trustworthy way of looking at the world—and at ourselves in the world.

But looking at the world through the prism of modern science has not been a simple matter. Until recently, science gave a fragmented picture of the world, conveyed through seemingly independent disciplinary compartments. Scientists have found it difficult to tell what connects the physical universe to the living world, the living world to the world of society, and the world of society to the domains of mind and culture. This is now changing; at the leading edge of the sciences ever more researchers are searching for a more integrated, unitary world picture. This is true especially of physicists, who are intensely at work creating "grand unified theories" and "super-grand unified theories." These GUTs and super-GUTs relate together the fundamental fields and forces of nature in a logical and coherent theoretical scheme, suggesting that they had common origins.

A particularly ambitious endeavor has surfaced in quantum physics in recent years: the attempt to create a theory of everything. This project is based on string and superstring theories (so called because in these theories elementary particles are viewed as vibrating filaments or strings). The TOEs being developed use sophisticated mathematics and multidimensional spaces in an attempt to produce a single master equation that could account for all the laws of the universe.

BACKGROUND BRIEF
THE PHYSICISTS' THEORIES OF EVERYTHING

The theories of everything that are being researched and developed by theoretical physicists endeavor to achieve what Einstein once called "reading the mind of God." He said that if we could bring together all the laws of physical nature into a consistent set of equations, we could explain all the features of the universe on the basis of those equations;

that would be tantamount to reading the mind of God.

Einstein's own attempt took the form of a unified field theory. Although he pursued this ambitious quest until his death in 1955, he did not find the simple and powerful equation that would explain all physical phenomena in a logically consistent form.

The way Einstein tried to achieve his objective was by considering all phenomena of interest to physics as the interaction of continuous fields. We now know that his failure was due to the disregard of the fields and forces that operate at the microphysical level of reality: these fields (the weak and the strong nuclear forces) are central to quantum mechanics, but not to relativity theory.

A different approach has been adopted today by the majority of theoretical physicists: they take quanta—the discontinuous aspect of physical reality—as basic. But the physical nature of quanta is reinterpreted: they are no longer discrete matter-energy particles but rather vibrating one-dimensional filaments: "strings" and "superstrings." Physicists try to link all the laws of physics as the vibration of superstrings in a higher dimensional space. They see each particle as a string that makes its own "music" together with all other particles. Cosmically, entire stars and galaxies vibrate together, as, in the final analysis, does the whole universe. The physicists' challenge is to come up with an equation that shows how one vibration relates to another, so that they can all be expressed consistently in a single super-equation. This equation would decode the encompassing music that is the vastest and most fundamental harmony of the cosmos.

At the time of writing, a string-theory-based TOE remains an ambition and a hope: nobody has come up with the super-equation that could express the harmony of the physical universe in a formula as simple and basic as Einstein's original $E = mc^2$. Indeed there are so many problems with it that ever more physicists claim that a fundamentally new concept may be needed to make progress. The equations of string theory call for multiple dimensions to pan out; not even a four-dimensional space-time suffices. Initially the theory required up to twenty dimensions to relate all the vibrations together in a consistent theory, but now it appears that "only" ten or eleven dimensions would do, provided that the vibrations occur in a higher-dimensional "hyperspace." Moreover string theory requires an existing framework of space and time for its strings, but it cannot show how space and time would be generated. Even more vexing is that the theory has so many possible solutions—of the order of 10^{500}—that it becomes a mystery why our universe is the way it is (seeing that each solution would produce a different universe).

Physicists who hope to salvage string theory have brought forward various hypotheses. It could be that all possible universes coexist, although we live in just one of them. Or, it could also be that our universe has a multitude of different faces but we perceive only the one familiar face. These are among a spate of hypotheses put forward by theoretical physicists who are intent on demonstrating that string theories have some measure of realism; they are theories of the real world. But none of them is satisfactory, and some critics, among them Peter Woit and Lee Smolin, are ready to bury string theory.

Smolin is one of the founders of the theory of loop quantum gravity, according to which space is a network of nodes that interconnects all points. The theory explains how space and time are generated and also accounts for "action-at-a-distance," that is, for the strange "entanglement" that underlies the phenomenon known as nonlocality, which we will explore in chapter three.

Whether physicists will be able to come up with a working theory of everything is evidently in doubt. It is clear, however, that even if the current efforts were to be crowned with success, success would not crown the creation of a genuine TOE. At the most, physicists would come up with a physical TOE—with a theory that is not a theory of *every* thing, only of every *physical* thing. A genuine TOE would include more than the mathematical formulas that give a unified expression to the phenomena studied in this branch of quantum physics. There is more to the universe than vibrating strings and related quantum events. Life, mind, culture, and consciousness are part of the world's reality, and a genuine theory of everything would take them into account as well.

Ken Wilber, who wrote a book with the title *A Theory of Everything*, agrees: he speaks of the "integral vision" conveyed by a genuine TOE. However, he does not offer such a theory; he mainly discusses what it would be like, describing it in reference to the evolution of culture and consciousness—and to his own theories. An actual, science-based integral theory of everything is yet to be created.

APPROACHES TO A GENUINE TOE

A genuine TOE *can* be created. Although it is beyond the string and superstring theories in the framework of which physicists attempt to formulate their own super-theory, it is well within the scope of science itself. Indeed, the enterprise of creating a genuine TOE—an I-TOE—is simpler than the attempt to create a physical TOE. As we have seen, physical TOEs endeavor to relate together all the laws of physics in a single formula—laws that govern interactions among particles and atoms, stars and galaxies: many already complex entities with complex interrelations. It is simpler, and more sensible, to look for the basic laws and processes that *give rise to* these entities, and to their interrelations.

The computer simulation of complex structures demonstrates that complexity is generated, and can be explained, by basic and relatively simple starting conditions. As John von Neumann's cellular automata theory has shown, it is enough to identify the basic constituents of a system and give the rules—the algorithms—that govern their behavior. (This is the basis of all computer simulations: the modelers tell the computer what to do at each step as the modeling process unfolds, and the computer does the rest.) A finite and surprisingly simple set of basic elements governed by a small set of algorithms can generate great and seemingly incomprehensible complexity merely by allowing the process to unfold in time. A set of rules informing a set of elements initiates a process that orders and organizes the elements, so that they create more and more complex structures and interrelations.

When we try to create a genuine I-TOE, we can proceed in an analogous way. We can start with the basic kind of things, the things that generate other things without being generated by them. Then we must state the simplest possible set of rules that can generate the more complex things. In principle we should then be able to explain how every "thing" in the world has come to be.

Beyond string and superstring theories, there are theories and concepts in the new physics through which this ambitious enterprise can be attempted. Using the findings of cutting-edge particle and field theories, we can identify the foundation that generates all things without itself being generated by other things. This foundation, we shall see, is the virtual-energy sea known as the *quantum vacuum*. We can also draw on a large repertory of rules—the laws of nature—that tell us how the basic elements of reality—the particles known as quanta—evolve into complex things in interaction with their cosmic foundation.

However, we must add a new element to achieve a genuine I-TOE. The currently known laws by which the existing things of the world are generated from the quantum vacuum are laws of interaction based on the transfer and transformation of *energy*. These laws turn out to be adequate to explain how real things—in the form of particle-antiparticle pairs—are generated in and emerge out of the quantum vacuum. But they do not adequately explain why an excess of particles over antiparticles was generated in the Big Bang, nor does it tell us how in the course of cosmic eons the surviving particles became structured into more and more complex things: into galaxies and stars, atoms and molecules, and—on suitable planetary surfaces—into macromolecules, cells, organisms, societies, ecologies, and entire biospheres.

In order to account for the presence of a significant number of particles in the universe (of "matter" as opposed to "anti-matter"), and for the ongoing, if by no means smooth and linear, evolution of the existing things, we need to recognize the presence of a factor that is neither matter nor energy. The importance of this factor is now acknowledged not only in the human and the social sciences, but also in the physical and the life sciences. It is *information*—information as a real and effective factor setting the parameters of the universe at its birth, and thereafter governing the evolution of its basic elements into complex systems.

Most of us think of information as data or what a person knows. But the reach of information is deeper than this. Physical and life scientists are discovering that information extends far beyond the mind of an individual person, or even all persons taken together. It is an inherent aspect of both physical and biological nature. The great physicist David Bohm called it "in-formation," meaning a process that actually "forms" the recipient. This is the concept we shall adopt here.

In-formation is not a human artifact, not something we produce by writing, calculating, speaking, and messaging. As ancient sages knew, and as scientists are now rediscovering, in-formation is present in the world independent of human volition and action and is a decisive factor in the evolution of the things that furnish the real world. The basis for creating a genuine I-TOE is the recognition that "in-formation" is a fundamental factor in nature.

TWO

On Puzzles and Fables

Drivers of the Next Paradigm Shift in Science

We begin our search for a genuine I-TOE by reviewing the factors that drive science toward a new paradigm. The key drivers are the puzzles that crop up and accumulate in the course of scientific investigation: anomalies that the current paradigm cannot clarify. This impels the community of scientists to search for new ways of approaching the anomalous phenomena. Their exploratory probes (we shall call them "science fables") bring to the surface many new ideas. Some of the ideas may harbor the key concepts that will lead scientists to a new paradigm: to a paradigm that could clear up the puzzles and anomalies and provide the foundation for a genuine I-TOE.

Leading scientists wish to extend and deepen their understanding of the segment of reality they investigate. They understand more and more of the pertinent part or aspect of reality, but they cannot inspect any part or aspect directly; they can only comprehend it through concepts built into hypotheses and theories. Concepts, hypotheses, and theories are not eternally valid—they are fallible. Indeed, the mark of a truly scientific theory, according to philosopher of science Sir Karl Popper, is its "falsifiability." Theories are falsified when the predictions made on their basis are not borne out by observations. In that case the observations are "anomalous," and the theory in question is either considered false and is abandoned, or is admitted to be in need of revision.

The falsification of theories is the engine of real progress in science. When everything works, there can still be progress, but it is piecemeal, the refinement of the accepted theory to match the new observations. Real progress occurs when this is not possible. Then the point is sooner or later reached when—instead of trying to revise the established theories—scientists prefer to look for a simpler and more insightful theory. The way is open to fundamental theory innovation: to a *paradigm-shift*.

A paradigm-shift is driven by the accumulation of observations that do not fit the accepted theories and cannot be made to fit by the mere extension of those theories. The stage is set for a new and more adequate scientific paradigm. The challenge is to find the fundamental, and fundamentally new, concepts that form the substance of the new paradigm.

There are stringent requirements on a scientific paradigm. A theory based on it must enable scientists to explain all the findings covered by the previous theory and must also explain the anomalous observations. It must integrate all the relevant facts in a simpler and yet more encompassing and powerful concept. This is what Einstein did at the turn of the twentieth century when he stopped looking for solutions to the puzzling behavior of light in the framework of Newtonian physics and created instead a new concept of physical reality: the theory of relativity. As he himself said, one cannot solve a problem with the same kind of thinking that gave rise to that problem. In a surprisingly short time, the bulk of the physics community abandoned the classical physics founded by Newton and embraced Einstein's revolutionary concept in its place.

In the first decade of the twentieth century, science underwent a basic "paradigm shift." Now, in the first decade of the twenty-first century, puzzles and anomalies are accumulating again, and the scientific community faces another paradigm shift, just as fundamental as the revolution that shifted science from the mechanistic world of Newton to the relativistic universe of Einstein.

The current paradigm shift has been brewing in the avant-garde circles of science for some time. Scientific revolutions are not instant-fit processes, with a new theory clicking into place all at once. They may be rapid, as in the case of Einstein's theory, or more protracted, as the shift from the classical Darwinian theory to a more embracing post-Darwinian conception in biology, for example.

Before the incipient revolutions are consolidated, the sciences affected by the anomalies go through a period of turbulence. Mainstream scientists defend the established theories, while maverick scientists at the cutting edge explore alternatives. The latter come up with new, sometimes radically different ideas that look at the same phenomena the mainstream scientists look at but see them differently. For a time, the alternative conceptions—initially in the form of working hypotheses—seem strange if not actually fantastic. They are something like fables, dreamt up by imaginative investigators. Yet they are not the work of untrammeled imagination. The "fables" of serious investigators are based on rigorous reasoning, bringing together what is already known about the segment of the world researched in a given discipline with what is as yet puzzling about it. They are not ordinary fables, but "science fables"—reasoned hypotheses that are testable and hence capable of being confirmed or shown to be false by observation and experiment.

Investigating the anomalies that crop up in observation and experiment and coming up with the testable fables that could account for them make up the nuts and bolts of fundamental research in science. If the anomalies persist despite the best efforts of mainstream scientists, and if one or another of the science fables advanced by maverick investigators gives a simpler and more logical explanation, a critical mass of scientists (mostly young ones) stops standing by the old paradigm.

That is the beginning of a paradigm shift. A concept that was until then a fable starts to be recognized as a valid scientific theory.

There are countless examples of successful as well as of failed fables in the history of science. Confirmed fables—presently valid even if not eternally true scientific theories—include Charles Darwin's concept that all living species descended from common ancestors and Alan Guth and Andrei Linde's hypothesis that the universe originated in a superfast "inflation" following its explosive birthing in the Big Bang. Failed fables—those that turn out not to be an exact, or at any rate the best, explanation of the pertinent phenomena—include Hans Driesch's notion that the evolution of life follows a preestablished plan in a goal-guided process called entelechy, and Einstein's hypothesis that an additional physical force, called the cosmological constant, keeps the universe from collapsing under the pull of gravitation. (Interestingly, as we shall see, some of these assessments are being questioned today: it may be that Guth's and Linde's "inflation theory" will be replaced by the more encompassing concept of a cyclical universe, and that Einstein's cosmological constant was not mistaken after all . . .)

A SAMPLING OF CURRENT SCIENCE FABLES

Here, by way of example, are three imaginative working hypotheses—"science fables"—put forward by highly respected scientists. All three have received serious attention in the scientific community, even though they are mind-boggling as descriptions of the real world.

10^{100} UNIVERSES

In 1955 the physicist Hugh Everett advanced a fabulous explanation of the quantum world (it was subsequently the basis for *Timeline,* one of Michael Crichton's best-selling novels). Everett's "parallel universes hypothesis" refers to a puzzling finding in quantum physics: that as long as a particle is not observed, measured, or interacted with in any way, it is in a curious state that is the superposition of all its possible states. When, however, the particle is observed, measured, or subjected to an interaction, this state of superposition becomes resolved: the particle is then in a single state only, like any "ordinary" thing. Because the state of superposition is described in a complex wave function associated with the name of Erwin Schrödinger, when the superposed state resolves it is said that the Schrödinger wave function "collapses."

The rub is that there is no way to tell which of its many possible "virtual states" the particle will then occupy. The particle's choice seems to be indeterminate—entirely independent of the conditions that trigger the wave function's collapse. Everett's hypothesis claims that the indeterminacy of the wave function's collapse does not reflect actual conditions in the world. There is no indeterminacy involved here: each virtual state selected by the particle is deterministic in itself—it simply takes place in a world of its own!

This is how the collapse would occur: When a quantum is measured, there are a number of possibilities, each of which is associated with an observer or a measuring device. We perceive only one of these possibilities in a seemingly random process of selection. But, according to Everett, the selection is not random, for it does not take place in the first place: all possible states of the quantum are realized every time it is measured or observed; they are just not realized in the same world. The many possible states of the quantum are realized in as many universes.

Suppose that when it is measured, a quantum such as an electron has a 50 percent probability of going up and a 50 percent probability of going down. Then we do not have just one universe in which the quantum has a 50/50 probability of going up or going down, but two parallel universes. In one of the universes the electron is actually going up and in the other it is actually going down. We also have an observer or a measuring instrument in each of these universes. The two outcomes exist simultaneously in the two universes, and so do the observers or measuring instruments.

Of course, when a particle's multiple superposed states resolve into a single state there are not just two, but a vast number of possible virtual states that this particle can occupy. Thus a vast number of universes must exist—perhaps of the order of 10^{100}—complete with observers and measuring instruments.

THE OBSERVER-CREATED UNIVERSE

If there are 10^{100} or even 10^{500} universes, and given that in all but a handful of them life could never arise, how is it that we live in a universe where life, even complex forms of life, could evolve? Is this pure serendipity? Many science fables address this question, including the anthropic cosmological principle that claims our observation of this universe has something to do with this fortunate state of affairs. Recently Stephen Hawking of Cambridge University and Thomas Hertog of CERN (the European Nuclear Research Organization) came up with a related, mathematically sophisticated answer. According to their "observer-created universe" theory, rather than individual universes branching off in the course of time and existing on their own (as string theory would suggest), every possible universe exists simultaneously, in a state of superposition. Our being in this universe selects the path that leads to this particular universe from among the other paths leading to all the other universes; the rest of the paths cancel out. Thus in their theory the causal chain of events is reversed: the present determines the past. This would not be possible if the universe had a definite initial state, for out of that unique state a unique history would follow. But, Hawking and Hertog claim, there is no definite initial state for the universe, no starting point: that "boundary" simply does not exist.

THE HOLOGRAPHIC UNIVERSE

This science fable claims that the entire universe is a hologram —or, at least, that it can be treated as such. (In a hologram, as we shall discuss later, a pattern in two dimensions generates an image in three dimensions.) All the information that constitutes the universe is said to be stored on its periphery, which is a two-dimensional surface. This two-dimensional information reappears inside the universe in three dimensions. We see the universe in three dimensions even though what makes it what it is, is a two-dimensional field of information. Why is this seemingly outlandish idea the subject of intense discussion and research?

The problem the holographic universe concept addresses comes from thermodynamics. According to its solidly established second law, disorder can never decrease in any closed system. This means that disorder cannot decrease in the universe as a whole because when we take the cosmos in its totality, it is a closed system: there is no "outside" and hence nothing to which it could be open. That disorder cannot decrease means that order—which can be represented as information—cannot increase. According to quantum theory, the information that creates or maintains order must be constant; it cannot increase, and it also cannot decrease.

But what happens to information when matter collapses into black holes? It would seem that black holes wipe out the information contained in matter. This, however, would fly in the face of quantum theory. In response to this riddle, Stephen Hawking, together with Jacob Bekenstein, then of Princeton University, worked out that disorder in a black hole is proportional to its surface area. Within the black hole there is a great deal more room for order and information than at its surface. In a single cubic centimeter, for example, there is room for 10^{99} Planck volumes inside, but room for only 10^{66} bits of information on the surface (a Planck volume is an almost inconceivably small space bounded by sides measuring 10^{-35} meter). Leonard Susskind of Stanford University and Gerard 't Hooft of the University of Utrecht came up with the idea that information inside the black hole is not lost—it is stored holographically on its surface.

The mathematics of holograms found unexpected application in 1998, when Juan Maldacena, then at Harvard University, tried to account for string theory under conditions of quantum gravity. Maldacena found that it is easier to deal with strings in five-dimensional spaces than in four dimensions. (We experience space in three dimensions: two planes along the surface and one up and down. A fourth dimension would be in a direction perpendicular to these, but this dimension cannot be experienced. Mathematicians can add any number of further dimensions, further and further removed from the world of experience.) The solution seemed evident: assume that the five-dimensional space inside the black hole is really a hologram of a four-dimensional pattern on its surface. One can then do the calculations in the more manageable five dimensions while dealing with a space of four dimensions.

Would a dimensional reduction work for the universe as a whole? As we have seen, string theorists are struggling with many extra dimensions, having discovered that three-dimensional space is not enough to accomplish their quest to relate the vibrations of the various strings in the universe in a single master-equation. The holographic principle would help, for the entire universe could then be considered a many-dimensional hologram, conserved in a smaller number of dimensions on its periphery.

The holographic principle may make string theory's calculations easier, but it makes fabulous assumptions about the nature of the world. Even Gerard 't Hooft, who was one of the originators of this principle, changed his mind about its cogency. Rather than a "principle," he said, in this context holography is actually a "problem." Perhaps, he speculated, quantum gravity could be derived from a deeper principle that does not obey quantum mechanics.

In periods of scientific revolution, when the established paradigm is under pressure, many science fables are put forward but not all pan out. Theoreticians proceed on the assumption that, as Galileo said, "the book of nature is written in the language of mathematics" and forget that not everything in the language of mathematics has a place in the book of nature. As a result many mathematically sophisticated fables remain just that—fables. Others, however, harbor the seeds of significant scientific advance. Initially, nobody knows for sure which of the seeds will grow and bear fruit. The field is in ferment, in a state of creative chaos.

In a remarkable variety of scientific disciplines this is the case today. A growing number of anomalous phenomena are coming to light in physical cosmology, in quantum physics, in evolutionary and quantum biology, and in the new field of consciousness research. They create growing uncertainties and induce open-minded scientists to look beyond the bounds of the established theories. While conservative investigators insist that the only ideas that can be considered scientific are those published in established science journals and reproduced in standard textbooks, cutting-edge researchers look for fundamentally new concepts, including some that were considered beyond the pale of their discipline but a few years ago.

In more and more disciplines the world is turning more and more fabulous. It is furnished with dark matter, dark energy, and multidimensional spaces in cosmology, with particles that are instantly connected throughout space-time by deeper levels of reality in quantum physics, with living matter that exhibits the coherence of quanta in biology, and with space- and time-independent transpersonal connections in consciousness research—to mention only some of the already validated science fables, now considered bona fide theories.

THREE

A Concise Catalog of the Puzzles of Coherence

We pursue our search for the I-TOE with a catalog of the findings that puzzle scientists today. Evidently, this catalog cannot cover all the puzzles that crop up in the different fields of scientific investigation. It does cover, however, a variety of puzzles that is significant in itself, and crops up surprisingly often in a variety of fields. These are puzzles of *coherence*. Intended is not the ordinary, garden variety of coherence, but the remarkable variety where the parts of the coherent system are so finely adjusted to each other that a change in any one of them introduces change in all the others. Moreover the changes propagate through the system quasi-instantly, and are enduring. It is as if the parts of the system are "nonlocal"—not limited to just where they are, but are in some way everywhere throughout the system. In this chapter we examine how this strange form of coherence appears in the physical world, in the living world, and in the world of consciousness.*

---

*The ideas and findings outlined here and in the next chapters are presented in a more detailed but also more technical form in Ervin Laszlo, *The Connectivity Hypothesis: Foundations of an Integral Science of Quantum, Cosmos, Life, and Consciousness* (Albany: State University of New York Press, 2003).

THE PUZZLES OF COHERENCE
IN QUANTUM PHYSICS

Coherence is a well-known phenomenon in physics: in its ordinary form, it refers to light as being composed of waves that have a constant difference in phase. Coherence means that phase relations remain constant and processes and rhythms are harmonized. Ordinary light sources are coherent over a few meters; lasers, microwaves, and other technological light sources remain coherent for considerably greater distances. But the kind of coherence discovered today is more complex and significant than the standard form. It indicates a quasi-instant correlation among the parts or elements of a system, whether that system is an atom, an organism, or a galaxy. All parts of a system of such coherence are so correlated that what happens to one part also happens to the other parts.

Coherence of the "nonlocal" kind is just one of the surprising phenomena that came to light in the twentieth century. The world picture of quantum physics—the physics of the ultrasmall domain of physical reality—became strange beyond imagination. The findings indicate that the smallest identifiable units of matter, force, and light are not entirely "separate realities" but specific forms and bundles of underlying energy fields. Some of these "quanta" have matterlike properties, such as mass, gravitation, and inertia. Others have force-properties, making up the particles that convey effective interaction among matterlike quanta. Yet others have lightlike properties; they carry electromagnetic waves that include the visible spectrum. But none of the quanta are truly separate from one another, for—once having shared the same state—they remain interlinked no matter how far they may be from each other. And none behave as ordinary objects. They have both corpuscular and wave-properties, depending, it seems, on the way the experiments through which they are observed are set up. Moreover when one of their properties is measured, the others become unavailable to observation and measurement.

THE WEIRD WORLD OF THE QUANTUM

THE PRINCIPAL LANDMARK: THE ENTANGLED PARTICLE

- In their pristine state, quanta are not just in one place at one time: each single quantum is both "here" and "there"—and in a sense it is everywhere in space-time.

- Until they are observed or measured, quanta have no definite characteristics but instead exist simultaneously in several states at the same time. These states are not "real" but "virtual"—they are the states the quanta can assume when they are observed or measured. It is as if the observer, or the measuring instrument, fishes the quanta out of a sea of possibilities. When a quantum is pulled out of that sea, it becomes a real rather than a mere virtual beast—but we can never know in advance just which of the possible real beasts that it *could* become it actually *will* become. It appears to choose its real states on its own from among the virtual states available to it.

- Even when the quantum is in a real state, it does not allow us to observe and measure all the parameters of its state at the same time: when we measure one parameter (for example, position or energy), another becomes blurred (such as its speed of motion or the time of its observation).

- Quanta are highly sociable: once they share the same identical state, they remain linked no matter how far they travel from each other. When one of a pair of formerly connected quanta is subjected to an interaction (that is, when it is observed or measured), it chooses its own "real" state—and its twin also chooses its own state, but not freely: it chooses it according to the choice of the first twin. The second twin always chooses a complementary state, never the same as the first twin.

- Within a complex system (such as the whole setup of a physics experiment) quanta exhibit just as sociable behaviors. If we measure one of the quanta in the system, the others shift from a virtual to a real state as well. Even more remarkably, if we create an experimental situation where a given quantum can be individually measured, all the other quanta become "real" even if the experiment is *not* carried out. . . .

Classical mechanics, the physics of Isaac Newton, conveyed a comprehensible concept of physical reality. Newton's *Philosophiae Naturalis Principia Mathematica,* published in 1687, demonstrated with geometrical precision that material bodies move according to mathematically expressible rules on Earth, while planets rotate in accordance with Kepler's laws in the heavens. The motion of all things is rigorously determined by the conditions under which it is initiated, just as the motion of a pendulum is determined by its length and its initial displacement and the trajectory of a projectile by its launch angle and acceleration. With mathematical certainty Newton predicted the position of the planets, the motion of pendulums, the path of projectiles, and the motion of the "mass points" that in his physics are the ultimate building blocks of the universe.

Over a hundred years ago, the mechanistic, predictable world of Newton ran into trouble. With the splitting of the atom in the late nineteenth century and of the atomic nucleus in the early twentieth, more had been fragmented than a physical entity. The very foundation of natural science was shaken: the experiments of early-twentieth-century physics demolished the prevailing view that all of reality is built of blocks that are themselves not further divisible. Yet physicists could not put any comparably commonsensical concept in its place. The very notion of "matter" became problematic. The subatomic particles that emerged when atoms and atomic nuclei were fissioned did not behave like conventional solids: they had a mysterious interconnection known as "nonlocality," and a dual nature consisting of wavelike as well as corpuscle-like properties.

It turned out that the particles that make up the manifest aspect of reality are not little mass points, like tiny balls of matter, but waves; more exactly, standing waves. They are described in quantum physics by wave functions. All visible order in the universe is determined by the rules that govern the interference of these waves. The possible patterns of interference among the standing waves we know as atoms determines what kind of molecules the atoms can form and hence what kind of chemical systems can come about. The pattern of interference of molecules determines in turn the possible kinds of intermolecular interactions, including the complex interactions that form the basis of life.

The kinds of interactions that are possible are determined in turn by the order of virtual states. As just remarked, every particle, every atom, and every molecule possesses not only the state that it occupies when it is observed, but also states that are empty and hence are said to be "virtual." Virtual states are described by probability functions and bits of information. They become real when a particle, an atom, or a molecule "jumps" into them.

The set of virtual states into which a given particle, atom, or molecule can jump—unlike the jumps themselves—is not random. The order of a given particle's (or atom's or molecule's) set of virtual states controls the translational, vibrational, and rotational motion of that particle (or atom or molecule). This virtual-state order determines the movement of chemical systems across surfaces of potential energy by leading them from one conformal state to another—

from one kind of chemical or biochemical ensemble to another.

Every system that emerges in the manifest world represents a selection from among the set of virtual states that is available to it. There is a constant transformation from virtual into real states and also from real into virtual states. Quantum physical-chemist Lothar Schäfer describes this as "an incessant, restless dance" where the occupied states are constantly abandoned and become virtual, while the empty states become occupied and real. As he writes, "at the foundation of things transcendent (that is, virtual) order and real order are interlocked in an uninterrupted frantic embrace."

The mysterious interaction of real and virtual states in the physical world is compounded by another mystery: the constant, and seemingly space- and time-transcending connection between particles in the real state. The famous "EPR" experiment (the experiment originally suggested by Albert Einstein together with colleagues Boris Podolski and Nathan Rosen) demonstrates that particles that at one time shared the same identical state (the same system of coordinates) remain instantly and enduringly connected. Such connection extends to entire atoms: current "teleportation" experiments show that when one of a pair of correlated atoms is further correlated with a third atom, the quantum state of the third is instantly transferred ("beamed") to the other of the initially connected pair—no matter how far away that atom may be.

The remarkable fact emerging from this sea of quantum mystery is that particles, and the atoms constituted by particles, are not individual beasts. They are sociable entities, and under certain conditions they are so thoroughly "entangled" with each other that they are not just here or there, but in all measured places at the same time. Their nonlocality respects neither time nor space: it exists whether the distance that separates the particles and the atoms is measured in millimeters or in light-years, and whether the time that separates them consists of seconds or of millions of years.

QUANTUM NONLOCALITY:
THE REVOLUTIONARY EXPERIMENTS

THE EPR EXPERIMENT

The EPR experiment—the first of the revolutionary experiments that testify to the nonlocality of the microsphere of physical reality—was put forward by Albert Einstein with his colleagues Boris Podolski and Nathan Rosen in 1935. This "thought experiment" (so-called because at the time it could not be empirically tested) requires that we take two particles in a so-called singlet state, where their spins cancel out each other to yield a total spin of zero. We then allow the particles to separate and travel a finite distance. If we could then measure the spin states of both particles, we would know both states at the same time. Einstein believed that this would show that the strange limitation on what can be measured given by Heisenberg's principle of uncertainty does not hold; the theory on which it is based does not offer a complete description of physical reality.

When experimental apparatus sophisticated enough to test the possibility that Einstein is right was devised, it turned out that this is not exactly what happens. Suppose that we measure the spin state of one of the particles—particle A—along some direction, let us say the *z-axis* (the permissible spin states are "up" or "down" along *axes x, y,* and *z*). Let us say we find that this measurement shows the spin to be in the direction "up." Because the spins of the particles have to cancel each other, the spin of particle B must definitely be "down." But the particles are removed from each other; this requirement should not hold. Yet it does. Every measurement on one particle yields a complementary outcome in the measurement on the other. It appears that the measurement of particle A has an instantaneous effect on B, causing its spin-wave function to collapse into the complementary state. The measurement on A does not merely *reveal* an already established state of B: it actually *produces* that state.

An instantaneous effect propagates from A to B, conveying precise information on what is being measured. B "knows" when A is measured, for what parameter, and with what result, for it assumes its own state accordingly. A nonlocal connection links A and B, notwithstanding the distance that separates them.

Experiments performed in the 1980s by Alain Aspect and collaborators and repeated by Nicolas Gisin in 1997 show that the speed with which the effect is transmitted is mind-boggling. In Aspect's experiments the communication between particles twelve meters apart was estimated at less than one billionth of a second, about twenty times faster than the speed with which light travels in empty space, and in Gisin's experiment particles ten kilometers apart appeared to be in communication 20,000 times faster than the velocity of light, relativity theory's supposedly unbreakable speed barrier. The experiments also show that the connection between the particles is not transmitted by conventional means through the measuring apparatus; it is intrinsic to the particles themselves. The particles are "entangled": their correlation is not sensitive either to distance in space or to difference in time.

Subsequent experiments have involved more particles over ever-larger distances, without modifying these surprising results. It appears that separation does not divide particles from each other—otherwise a measurement on one would not produce an effect on the other. It is not even necessary that the particles should originate in the same quantum state. Experiments show that any two particles, be they electrons, neutrons, or photons, can originate at different points in space and in time; if they once come together within the same system of coordinates, that is enough for them to continue to act as part of the same quantum system even when they are separated . . .

THE TELEPORTATION EXPERIMENTS

Recent experiments show that a form of nonlocal connection known as "teleportation" exists not only between individual particles, but also between entire atoms. Teleportation has been experimentally proven since 1997 in regard to the quantum state of photons in light beams and the state of magnetic fields produced by clouds of atoms. In the spring of 2004 milestone experiments by two teams of physicists, one at the National Institute of Standards in Colorado and the other at the University of Innsbruck in Austria, demonstrated that the quantum state of entire atoms can be teleported by transporting the quantum bits ("qubits") that define the atoms. In the Colorado experiment led by M. D. Barrett, the ground state of beryllium ions was successfully teleported, and in the Innsbruck experiment headed by M. Riebe, the ground and metastable states of magnetically trapped calcium ions were teleported. The physicists achieved teleportation of a remarkably high fidelity—78 percent by the Colorado team and 75 percent by the Innsbruck team. The Colorado and Innsbruck scientists used different techniques but followed the same basic protocol.

First two charged atoms (ions), labeled A and B, are "entangled," creating the instant link that is also observed in the EPR experiment. Then a third atom, labeled P, is prepared by encoding in it the coherently superposed quantum state that is to be teleported. Then A, one of the entangled ions, is measured together with the prepared atom P. At that point the internal quantum state of B transforms: it assumes the exact state that was encoded in P! It appears that the quantum state of P has been "teleported" to B.

Although the experiments involve complex procedures, the real-world process they demonstrate is relatively straightforward. When A and P are measured together, the preexisting nonlocal connection between A and B creates a nonlocal transfer of state from P to B. In the EPR experiment, one of a pair of entangled particles "in-forms" the other of its measured state; similarly, in teleportation experiments, the measurement of one of a pair of entangled ions together with a third ion encodes the state of the latter in the other twin. Because the process destroys the superposed quantum state of A and re-creates it in P, it recalls science fiction's idea of "beaming" an object from one place to another.

While beaming entire objects, not to mention people, is far beyond the current realm of possibilities, the equivalent process on the human level can be envisaged. In this "thought experiment" we take two persons who are emotionally close to each other, let us say Archie and Betty, young people deeply in love. We ask a third person, Petra, to concentrate on a given thought or image. We then create a deep "transpersonal" connection between Archie and Petra by having them pray or meditate together. If human-level teleportation works, at the very instant Archie and Petra enter their shared meditative state, the thought or image Petra has been concentrating on vanishes from her own mind, to reappear in the mind of Betty.

Teleportation experiments open vast, and for the time being more realistic, vistas. Before long physicists will find ways to beam qubits not just from one atom to another, but among a large number of particles simultaneously. This will lead to various technological innovations, including a new generation of superfast quantum computers. When a large number of entangled particles are distributed through the structure of a computer, "quantum teleportation" can create an instant transfer of information among them without requiring that they be wired together, or even be next to each other.

THE PUZZLES OF COHERENCE IN COSMOLOGY

Cosmology, a branch of the astronomical sciences, is in turbulence. The deeper the new high-powered instruments probe the far reaches of the universe, the more mysteries they uncover. For the most part, these mysteries have a common element: they exhibit a staggering coherence throughout the reaches of space and time.

THE SURPRISING WORLD
OF THE NEW COSMOLOGY

THE PRINCIPAL LANDMARK: THE COHERENT
AND COHERENTLY EVOLVING COSMOS

The universe is far more complex and coherent than anyone other than poets and mystics have dared to imagine.
A number of puzzling observations have cropped up:

- *The violation of charge and parity.* A universe born in
  the burst of energy known as a Big Bang should contain
  equal numbers of particles and antiparticles—matter and
  antimatter. But if that had been true of our universe, the
  colliding pairs of particles and antiparticles would have
  annihilated each other and space-time would be empty
  of anything resembling what we could call matter. Yet
  there is no parity between matter and antimatter in the
  universe: there is a sufficient surplus of matter to furnish
  the cosmic space with particles, atoms, stars, and galaxies. (This puzzle is known as a "CP violation," where C
  is "charge conjugation" and P is "parity inversion," such
  as is seen in a mirror reflection.)

- *The energy of "empty" space.* Even in the absence of
  matter, cosmic space is not empty; a number of fields
  occupy it with positive energy values. As we shall discuss these include the zero-point field or ZPF (so-called
  because in this field energies prove to be present even
  when all classical forms of energy vanish: at the absolute zero of temperature) as well as the currently much
  debated Higgs field. The precise value of the energy
  present in matter-free (i.e., "empty") space may be the

critical, and as yet unknown, factor that determines whether the universe will expand forever, contract and head toward a Big Crunch, or remain balanced at the razor's edge between expansion and contraction.

- *The accelerating expansion of the cosmos.* Distant galaxies pick up speed as they move away from each other. Yet they should be slowing down as gravitation brakes the force of the Big Bang that blew them apart.

- *The "missing mass" of the universe.* There is more gravitational pull in the cosmos than visible matter can account for—yet only matter is believed to have mass and thus to exert the force of gravitation. Even when cosmologists allow for a variety of "dark" (optically invisible) matter, there is still a great chunk of matter (and hence mass) missing.

- *The coherence of some cosmic ratios.* The mass of elementary particles, the number of particles, and the forces that exist between them are all mysteriously adjusted to favor certain ratios that recur again and again.

- *The "horizon problem."* The galaxies and other macrostructures of the universe evolve almost uniformly in all directions from Earth, even across distances so great that the structures could not have been connected by light, and hence could not have been correlated by signals carried by light (because, according to relativity theory, no signal can travel faster than light).

- *The fine-tuning of the universal constants.* The key parameters of the universe are amazingly fine tuned to produce not just recurring harmonic ratios, but also the—statistically extremely improbable—conditions under which life can emerge and evolve in the cosmos.

According to the most widely accepted cosmological model, the so-called Big-Bang theory, the universe originated in a stupendous explosion twelve to fifteen billion years ago. (The standard estimate has been 13.7 billion years, but in 2006 a research team led by Alceste Bonanos at the Carnegie Institution in Washington came up with a different figure: the universe, they claim, is 15.8 billion years old.) The Big Bang must have been an explosive instability in the quantum vacuum. A region of this vacuum—which was, and is, far from a real vacuum, that is, empty space—exploded, creating a fireball of staggering heat and density. In the first milliseconds it synthesized all the matter that now populates cosmic space. The particle-antiparticle pairs that emerged collided with and annihilated each other. But for some reason—which is not explained either by the Big Bang theory or by the celebrated "Standard Model of particle physics" (the theory that provides the mathematics of the state and interaction of particles) more matter than antimatter particles were created, and the excess matter-particles make up the furnishings of the universe.

After about 400,000 years the universe cooled enough that charged electrons and protons could combine to form hydrogen atoms. Most of the quanta of light (photons) escaped the hot plasma, and as a result space became transparent. Clumps of particles (predominantly hydrogen atoms) established themselves as distinct elements in the cosmos, and matter in these hydrogen clumps condensed under gravitational attraction. In the span of one billion years, the first galaxies were formed. Within the galaxies subsidiary clumps came about, they heated up as they became more dense, and ultimately they reached the temperature where nuclear chain reactions could get under way. The stars began to shine.

Until quite recently, the scenario of cosmic evolution seemed well established. Detailed measurements of the cosmic microwave background radiation—the presumed remnant of the Big Bang—testify that its variations derive from minute fluctuations within the cosmic fireball when our universe was less than one trillionth of a second "young" and are not distortions caused by radiation from stellar bodies.

But the standard cosmology of the Big Bang ("BB theory") is not as established now as it was a few years ago. A growing number of mysteries have been cropping up. First, there is the unexplained CP violation at the birth of the universe. Then BB theory has nothing to say about the mysterious force that pushes apart the galaxies. This repulsive force is known as the "cosmological constant" and its value is estimated on the basis of quantum physics. The classical version of Big Bang theory is silent on the score of dark matter and dark energy and hence cannot account for the observed shortfall of gravitational mass in space (the "missing mass" problem). And BB theory offers no explanation of the coherence of some basic cosmic ratios, or for the uniformity of macrostructures throughout cosmic space (the "horizon problem").

The phenomenon cosmologists call the "fine-tuning of the universal constants" is particularly vexing. The three dozen or more physical parameters of the universe are so precisely adjusted that together they create the highly improbable conditions under which life can emerge on Earth—and presumably on other planets as well—and evolve to progressively higher levels of complexity.

These are all puzzles of coherence, and they raise the possibility that this universe did not arise in the context of a random fluctuation of the underlying quantum vacuum. Instead, it may have been born in the womb of a prior "meta-universe": a Metaverse. (The term *meta* comes from classical Greek, signifying "behind" or "beyond," in this case meaning a vaster, more fundamental universe that is behind or beyond the universe we inhabit.)

The existence of a vaster, perhaps infinite universe is underscored by the astonishing finding that no matter how far and wide high-powered telescopes range in the universe, they find galaxy after galaxy, even in "black regions" of the sky where no galaxies or stars of any kind were believed to exist. This is a very different picture from that which reigned in astronomy but a hundred years ago. At that time, and until the 1920s, it was thought that the Milky Way is all there is to the universe: where the Milky Way ends, space itself ends. Not only do we know today that the Milky Way—"our galaxy"—is but one among billions of other galaxies in "our universe," but we are also beginning to recognize that the boundaries of "our universe" are not the boundaries of "*the* universe." The cosmos may be infinite in time, and perhaps also in space—it is certainly vaster by several magnitudes than any cosmologist would have dared to envisage just a few decades ago.

A number of physical cosmologies offer quantitatively elaborated accounts of how our universe could have arisen in the framework of the Metaverse. Such cosmologies harbor the promise of overcoming the puzzles posed by the coherence of this universe, including the mind-blowing serendipity that its physical constants are so finely adjusted that we can be here to ask questions about them. This has no credible explanation in a one-shot, single-cycle universe, for there the vacuum fluctuations that set the parameters of the emerging universe would have to have been randomly selected: there was "nothing there" that could have biased the serendipity of this selection. Yet a random selection from among all the possible fluctuations in the chaos of a turbulent primordial vacuum is astronomically unlikely to have led to a universe where living organisms and other complex and coherent phenomena arise and evolve—or even to a universe in which there is a significant surplus of matter over antimatter.

The coherence of our universe tells us that all its stars and galaxies are connected in some way. And the astonishing fine-tuning of its physical constants suggests that at its birth the vacuum in which our universe emerged was not randomly structured. A previous universe is likely to have informed the birth of our universe, much as the genetic code of our parents informed the conception and growth of the embryo that grew into what we are today.

SOME CURRENT METAVERSE HYPOTHESES

A widely discussed hypothesis advanced by the Princeton physicist John Wheeler claims that the expansion of the universe will come to an end, and ultimately the universe will collapse back on itself. Following this "Big Crunch" it could explode again, giving rise to another universe. In the quantum uncertainties that dominate the supercrunched state, almost infinite possibilities exist for universe creation. This could account for the fine-tuned features of our universe since, given a sufficiently large number of successive universe-creating oscillations, even the improbable fine-tuning of a universe such as ours has a chance of coming about.

It is also possible that many universes come into being at the same time. This, in turn, is the case if the explosion that gave rise to them was reticular—made up of a number of individual regions. According to Andrei Linde's "inflation theory," the Big Bang had distinct regions, much like a soap bubble in which smaller bubbles are stuck together. As such a bubble is blown up, the smaller bubbles become separated and each forms a distinct bubble of its own. The bubble universes percolate outward and follow their own evolutionary destiny. Each bubble universe hits on its own set of physical constants, and these may be very different from those of other universes. For example, in some universes gravity may be so strong that they recollapse almost instantly; in others, gravity may be so weak that no stars could form. We happen to live in a bubble with physical constants that permit the evolution of complex systems, including human beings.

New universes could also be created inside black holes. The extreme high densities of these space-time regions represent singularities where the known laws of physics do not apply. Stephen Hawking and Alan Guth suggested that under these conditions the black hole's region of space-time detaches itself from the rest and expands to create a universe of its own. The black hole of one universe may be the "white hole" of another: the Big Bang that creates it.

In another cosmology baby universes are periodically created in bursts similar to that which brought forth our own universe. The QSSC (Quasi-Steady State Cosmology) advanced by Fred Hoyle together with George Burbidge and J. V. Narlikar postulates that such "matter-creating events" are interspersed throughout the meta-universe. Matter-creating events come about in the strong gravitational fields associated with dense aggregates of preexisting matter, such as in the nuclei of galaxies. The most recent burst occurred some fourteen billion years ago, in overall agreement with the age of our own universe.

Yet another Metaverse hypothesis was developed by Ilya Prigogine, J. Geheniau, E. Gunzig, and P. Nardone. Their theory agrees with the QSSC in suggesting that major matter-creating bursts similar to our Big Bang occur from time to time. The large-scale geometry of space-time creates a reservoir of "negative energy" (which is the energy required to lift a body away from the direction of its gravitational pull); from this reservoir, gravitating matter extracts positive energy. Thus gravitation is at the root of the ongoing synthesis of matter: it produces a perpetual matter-creating mill. The more particles are generated, the more negative energy is produced and then transferred as positive energy to the synthesis of still more particles. Given that the quantum vacuum is unstable in the presence of gravitational interaction, matter and vacuum form a self-generating feedback loop. A critical matter triggered instability causes the vacuum to transit to the inflationary mode, and that mode marks the beginning of a new era of matter synthesis.

A more recent cosmology is the work of Paul J. Steinhardt of Princeton and Neil Turok of Cambridge. It accounts for all the facts accounted for by the Big Bang theory and also gives an explanation of the puzzling acceleration of the expansion of distant galaxies. According to Steinhardt and Turok, the universe undergoes an endless sequence of cosmic epochs, each of which begins with a "Bang" and ends in a "Crunch." Each cycle includes a period of gradual and then further accelerating expansion, followed by reversal and the beginning of an epoch of contraction. They estimate that at present we are about fourteen billion years into the current cycle and at the beginning of a trillion-year period of accelerated expansion. Ultimately our universe (more exactly, our cycle of the universe) will achieve a condition where space is homogeneous and no longer curved. The next cycle will then begin.

THE PUZZLES OF COHERENCE IN BIOLOGY

The superlarge as well as the ultrasmall domains of physical reality turn out to be amazingly coherent. But the world in its everyday dimension is more reasonable. Here things occupy but one state at a time and are either here or there and not in both places simultaneously. This, at any rate, is the commonsense assumption, and on first sight it makes sense. The living organism is made up of cells, which are made up of molecules, which in turn are made up of atoms, made up of particles. The classical view insists that, even if particles themselves are weird, the whole made up of them is a classical object: the quantum indeterminacies are canceled out at the macroscale. But this is not—or at any rate not entirely—the case. Instant, multidimensional connections have come to light between the parts of a living organism, and even between organisms and their environments.

Cutting-edge research in quantum biology finds that atoms and molecules within organisms, and entire organisms and their environments, are nearly as "entangled" with each other as microparticles that originate in the same quantum state.

THE UNEXPECTED WORLD OF POST-DARWINIAN BIOLOGY

THE PRINCIPAL LANDMARK:
THE ULTRA-COHERENT ORGANISM

- The living organism is extraordinarily coherent: all its parts are multidimensionally, dynamically, and almost instantly connected with all other parts. What happens to one cell or organ also happens in some way to all other cells and organs—a connection that recalls (and in fact suggests) the kind of "entanglement" that characterizes the behavior of quanta in the microdomain.
- The organism is also coherent with the world around it: what happens in the external milieu of the organism

is reflected in some ways in its internal milieu. Thanks to this coherence, the organism can evolve in tune with its environment. The genetic makeup of even a simple organism is so complex, and its "fit" to the milieu so delicate, that in the absence of such "inside-outside tuning," living species could not mutate into viable forms before being eliminated by natural selection. That our world is not populated solely by the simplest kinds of organisms, such as bacteria and blue-green algae, is due in the last analysis to the kind of "entanglement" that exists among genes, organisms, organic species, and their niches within the biosphere.

That the living organism is coherent as a whole is not surprising—what is surprising is the *degree* and *form* of its coherence. The organism's coherence goes beyond the coherence of a biochemical system; in some respects it attains the coherence of a quantum system.

Evidently, if living organisms are not to succumb to the constraints of the physical world, their component parts and organs must be precisely yet flexibly correlated with each other. Without such correlation, physical processes would soon break down the organization of the living state, bringing it closer to the inert state of thermal and chemical equilibrium in which life as we know it is impossible. Near-equilibrium systems are largely inert, incapable of sustaining processes such as metabolism and reproduction, essential to the living state. An organism is in thermal and chemical equilibrium only when it is dead. As long as it is living, it is in a state of *dynamic* equilibrium in which it stores energy and information and has them available to drive and direct its vital functions.*

On deeper analysis it turns out that dynamic equilibrium requires a very high degree of coherence: it calls for instantaneous long-range correlations throughout the organism. Simple collisions among neighboring molecules—mere billiard-ball push-impact relations among them—must be complemented by a network of instant communication that correlates all parts of the living system, even those that are distant from one another. Rare molecules, for example, are seldom contiguous, yet they find each other throughout the organism. There would not be sufficient time for this to occur by a random process of jiggling and mixing; the molecules need to locate and respond to each other specifically, even if they are distant. It is difficult to see how this could be achieved by mechanical or chemical connections among the organism's parts, even if it is transmitted by a nervous system that reads biochemical signals from genes through DNA, RNA, proteins, enzymes, and neural transmitters and activators.

In a complex organism the challenge of maintaining dynamic equilibrium is gigantic. The human body consists of some million billion cells, far more than stars in the Milky Way galaxy. Of this cell population, 600 billion are dying and the same number are regenerating every day—over 10 million cells per second. The average skin cell lives only for about two weeks; bone cells are renewed every three months. Every ninety seconds millions of antibodies are synthesized, each from about twelve hundred amino acids, and every hour 200 million erythrocytes are regenerated. There is no substance in the

---

*The difference between thermal/chemical and dynamic equilibrium can be illustrated in reference to the movement of a ball over a hilly landscape. When the ball is at the bottom of a valley, it is at rest; if any force moves it out of its position, it will roll back to it. This is similar to thermal and chemical, so-called thermodynamical, equilibrium. But when the ball is at the top of a hill, it will roll down unless it can dynamically balance itself in its unstable position. This balancing act is an instance of dynamic equilibrium.

body that is constant, though heart and brain cells endure longer than most. And the substances that coexist at a given time produce thousands of biochemical reactions in the body each and every second.

No matter how diverse the cells, organs, and organ systems of the organism, in essential respects they act as one. According to the experimental biophysicist Mae-Wan Ho they behave like a good jazz band, where every player responds immediately and spontaneously to the improvisations of the others. The super jazz band of an organism never ceases to play in a lifetime, expressing the harmonies and melodies of the individual organism with a recurring rhythm and beat but with endless variations. Always there is something new, something made up, as it goes along. It can change key, change tempo, or change tune, as the situation demands, spontaneously and without hesitation. There is structure, but the real art is in the endless improvisations, where each and every player, however small, enjoys maximum freedom of expression, while remaining perfectly in step with the whole.

The "music" of a higher organism ranges over more than seventy octaves. It is made up of the vibration of localized chemical bonds, the turning of molecular wheels, the beating of microcilia, the propagation of fluxes of electrons and protons, and the flowing of metabolites and ionic currents within and among cells through ten orders of spatial magnitude.

The level of coherence exhibited by organisms suggests that quantum-type processes take place in them. This is borne out by experiment. Organisms are known to respond even to extremely low frequency electromagnetic radiation, and to magnetic fields so weak that only the most sophisticated instruments can register them. But radiation below molecular dimensions could not affect molecular assemblies unless a large number of molecules were super-coherently linked among themselves. Such linkages could come about only if quantum processes complement the organism's biochemical processes. They do, and as a result the living organism is in some respects a "macroscopic quantum system."

Connection within the organism embraces the set of the organism's genes, the so-called genome. This is an anomaly for mainstream biology. According to classical Darwinism, the genome is insulated from the vicissitudes that befall the rest of the organism. There is a full and complete separation of the *germ line* (the genetic information handed down from parent to offspring) from the *soma* (the organism that expresses the genetic information). Darwinists claim that in the course of successive generations in the life of a species, the germ line varies randomly, unaffected by influences acting on the soma. Evolution proceeds by a selection from among the randomly created genetic variants according to the "fit" of the soma (the resulting organism) to its particular environment. If this were so, biological evolution would be the product of a twofold chance: the chance variation of the genome, and the chance fit of the resulting mutants to their environment. To cite the metaphor made popular by the Oxford biologist Richard Dawkins, evolution would occur through trial and error: it would be the work of a blind watchmaker.

However, the classical Darwinian tenet regarding the isolation of the genome is not correct. There are many ways that the genome is affected by what happens to the organism. Through the "epigenome" (an array of chemical markers and switches located along the double helix of the DNA) even the way the organism is nourished affects how particular genes work: whether they are switched "on" or "off." Then there are laboratory experiments that show that mechanical force and exposure to chemicals and to radiation can rearrange the sequence of genes, creating a genetic mutation. And there is indirect evidence furnished by the evolutionary history of life on Earth. It indicates that genome, organism, and environment form an integrated system where functionally autonomous parts are so correlated that the organism can survive, and can produce offspring that prove viable under conditions that would have been fatal to the parent. The evidence is indirect but cogent, for in the absence of such correlation, the probability that complex organisms could have evolved on Earth in the 600 million years that have elapsed since the late Cambrian era would be negligibly small.*

A direct connection between the genome and the soma is shown by laboratory experiments. The cell biologist A. Maniotis described an experiment where a mechanical force impressed on an external cellular membrane was transmitted to a cell nucleus, which produced a mutation almost instantly. The experimentalist Michael Lieber went further. His work demonstrated that mechanical force acting on the outer membrane of cells is but one variety of interaction that results in a genetic rearrangement: any stress coming from the environment, mechanical or not, triggers a global "hypermutation."

The genome proves to be dynamic and adaptive: when challenged it creates complex and practically instant rearrangements. When plants and insects are subjected to toxic substances, they often mutate their gene pool in precisely such a way that detoxifies the poisons and creates resistance to them. The "adaptive response" of the genome is also evident when electromagnetic or radioactive fields irradiate the organism: this, too, has a direct effect on the structure of its genes. In many

---

*The mathematical physicist Fred Hoyle illustrated the probabilities involved using the example of the task of unscrambling the colored faces of a Rubik's cube. (This is a cube of which the six faces are subdivided into three color-coded sections each. The colors can be ordered by twisting the individual segments.) Assume that a blind man is trying to order the faces of this cube. He is handicapped by not knowing whether any twist he gives the cube brings him closer to or farther from his goal of ordering the sections of the cube. He is obliged to work by random trial and error, with the result that his chances of achieving a simultaneous color matching of the six faces of the cube are in the range of 1:1 to $1:5 \times 10^{18}$. If the blind man is to work through all the possible moves at the rate of one move per second, he will need 5×10^{18} seconds. This, however, he could not do, for 5×10^{18} seconds is 126 billion years—almost ten times more than the age of our universe!

The situation changes dramatically if the blind man receives prompting in his efforts. If he receives a correct "yes" or "no" prompt at each move, the laws of probability show that he will unscramble the cube at an average of 120 moves. Working at the rate of one move per second, he will need not *126 billion* years to reach his goal, but *two minutes*.

cases the new arrangement shows up in the offspring. Experiments in Japan and the United States show that rats develop diabetes when a drug administered in the laboratory damages the insulin-producing cells of their pancreas. These diabetic rats produce offspring in which diabetes arises spontaneously. It appears that the alteration of the rats' body cells produces a rearrangement of their genes.

Even more striking are experiments in which particular genes of a strain of bacterium are rendered defective—for example, genes that enable bacteria to metabolize lactose. When the bacteria are fed a pure milk diet, some among them mutate back precisely those of their genes that enable them to metabolize it again. Given the complexity of the genome even of humble bacteria, this response is astronomically unlikely to occur purely by chance.

The German theoretician Marco Bischof summed up the insight emerging at the frontiers of the life sciences. "Quantum mechanics has established the primacy of the inseparable whole. For this reason," he said (and the emphasis is his), "the basis of the new biophysics must be the insight into the fundamental interconnectedness *within* the organism as well as *between* organisms, and that of the organism *with the environment.*"

THE PUZZLES OF COHERENCE IN CONSCIOUSNESS

Consciousness is the most intimately and immediately known fact of our experience. It accompanies us from birth, presumably until death. It is unique, and seems to belong uniquely to each of us. Yet "my" consciousness may not be solely and uniquely mine. The connections that bind "my" consciousness to the consciousness of others, well known to traditional—so-called primitive, but in fact in many respects highly sophisticated—peoples, are rediscovered today in controlled experiments with thought and image transference, and the effect of the mind of one individual on the mind and body of another.

THE TRANSPERSONAL
WORLD OF CONSCIOUSNESS

THE PRINCIPAL LANDMARK:
THE CONNECTEDNESS OF THE HUMAN MIND

- Native tribes seem able to communicate beyond the range of eye and ear. As shown by the customs, buildings, and artifacts of diverse peoples who live on different points of the globe, and may have lived at different times in history, entire cultures appear to share information among themselves, even though they are not in any known form of contact with each other.

- In the laboratory also modern people display a capacity for the spontaneous transference of impressions and images, especially when they are genetically or emotionally close to each other.

- Some images and ideas—universal symbols and archetypes—occur and recur in the culture of all civilizations, modern and ancient, whether or not their people have known each other or have even known of each other's existence.

- The mind of one person appears able to act on the brain and body of another. This faculty, known to traditional peoples, is verified today in controlled experiments and forms the basis of a new branch of medicine known as *telesomatic* or *nonlocal* medicine.

Current findings at the farther reaches of human consciousness recall Einstein's pronouncement half a century ago. "A human being" he said, "is part of the whole, called by us 'universe,' a part limited in time and space. He experiences his thoughts and feelings as something separate from the rest—a kind of optical delusion of his consciousness. This delusion is a sort of prison for us, restricting us to our personal decisions and to affection for a few persons nearest to us." While in the conservative view human communication and interaction is limited to our sensory channels (everything that is in the mind, it is said, must first have been in the eye or ear), leading psychologists, psychiatrists, and consciousness researchers are rediscovering what Einstein realized and ancient cultures have always known: that we are linked by more subtle and encompassing connections as well. In current scientific literature these connections are called *transpersonal*.

Traditional cultures did not regard transpersonal connections with distant peoples, tribes, or cultures as illusion, but modern societies do. The modern mind is not ready to accept anything as real that is not "manifest"—not literally "ready to hand" (*manus* being Latin for "hand"). Consequently, transpersonal connections are viewed as paranormal and admitted only under exceptional conditions.

One of the exceptions is "twin pain"—where one of a pair of twins senses the pain or trauma of the other. This phenomenon is accepted as real, for it is well documented. Guy Playfair, who wrote the book *Twin Telepathy,* noted that about 30 percent of twins experience telepathic interconnection. He cites a 1997 television program where the production team tested four pairs of identical twins. The brain waves, blood pressure, and galvanic skin response of the four pairs of twins were rigorously monitored. One of the unsuspecting twins in each pair was subjected to a loud alarm fitted to the back of the chair in which he or she was sitting. In three of the four pairs, the other twin registered the resulting shock, even through he or she was closeted some distance away in a separate and soundproof room. The successful pairs were used for the show that went live on the air, and they again showed the telepathic connection, although the receiving twin could not give an account of what it was that the other twin had experienced. The technical supervisor of the show concluded that the distant twins "certainly picked up something from somewhere."

Identical twins are only the top of the tree of bonded pairs. Some form of telepathy has been observed among all people who share a deep bond, such as mothers and children, lovers, long-term couples, even close friends. In these cases all but the most conservative psychologists are forced to recognize the existence of some transpersonal connection. But only exceptionally broad-minded psychologists admit that transpersonal connection includes the ability to transmit actual thoughts and images, and that it is given to many and perhaps to all people. Yet this is the finding of recent experiments.

The telepathic powers of people—their ability to carry out various forms of thought and image transference—is not just wishful thinking or a misreading of the results. A whole spectrum of experimental protocols has been developed, ranging from the noise-reduction procedure known as the Ganzfeld technique to the rigorous "distant mental influence on living systems" (DMILS) method. Explanations in terms of hidden sensory cues, machine bias, cheating by subjects, and experimenter incompetence or error have all been considered, but were found unable to account for a number of statistically significant results. It appears that even "normal" people possess "paranormal" powers.

THREE PIONEERING TRANSPERSONAL EXPERIMENTS

1. In the early 1970s a team of two physicists, Russell Targ and Harold Puthoff, undertook one of the first experiments in controlled transpersonal thought and image transference. Targ and Puthoff placed the "receiver" in a sealed, opaque, and electrically shielded chamber

and the "sender" in another room where he or she was subjected to bright flashes of light at regular intervals. The brain-wave patterns of both sender and receiver were registered on electroencephalograph (EEG) machines. As expected, the sender exhibited the rhythmic brain waves that normally accompany exposure to bright flashes of light. However, after a brief interval, the receiver also began to produce the same patterns, although he or she was not being directly exposed to the flashes and was not receiving ordinary sense-perceivable signals from the sender.

Targ and Puthoff also conducted experiments on remote viewing. In these tests, distances that precluded any form of sensory communication between them separated sender and receiver. At a site chosen at random, the sender acted as a "beacon" and the receiver tried to pick up what the sender saw. To document their impressions, receivers gave verbal descriptions, sometimes accompanied by sketches. Independent judges found the descriptions of the sketches matched the characteristics of the site that was actually seen by the sender on average 66 percent of the time.

2. The second series of pioneering experiments is the work of Jacobo Grinberg-Zylberbaum, of the National University of Mexico. He performed more than fifty experiments over five years on spontaneous communication among individual test subjects. He paired his subjects inside soundproof and electromagnetic radiation-proof "Faraday cages" and asked them to meditate together for twenty minutes. Then he placed them in separate Faraday cages where one subject was stimulated and

the other not. The stimulated subject received stimuli at random intervals in such a way that neither he (or she) nor the experimenter knew when they were applied. The subjects who were not stimulated remained relaxed, with eyes closed. They were asked to try to feel the presence of the partner without knowing anything about his or her stimulation.

Typically, a series of one hundred stimuli were applied—such as flashes of light, sounds, and short, intense, but not painful electric shocks to the index and ring fingers of the right hand. The electroencephalograph (EEG) brain-wave records of both subjects were then synchronized and examined for "normal" potentials evoked in the stimulated subject and "transferred" potentials in the nonstimulated one. Transferred potentials were not found in control situations where there was no stimulated subject, when a screen prevented the stimulated subject from perceiving the stimuli (such as light flashes), or when the two subjects did not previously interact. But during experimental situations with stimulated subjects and with prior contact among them, the transferred potentials appeared consistently in about 25 percent of the cases. A young couple, deeply in love, furnished a particularly poignant example. Their EEG patterns remained closely synchronized throughout the experiment, testifying that their report of feeling deep oneness was not an illusion.

In a limited way, Grinberg-Zylberbaum could also replicate his results. When one individual exhibited the transferred potentials in one experiment, he or she usually exhibited them in subsequent experiments also.

The results did not depend on spatial separation between senders and receivers—the transferred potentials appeared no matter how far or how near they were to each other.

3. The third experiment involves dowsing. It turns out that dowsers can often pinpoint the location of water veins with great accuracy. Dowsing rods as well as pendulums respond to the presence of underground water, magnetic fields, and even oil and other natural substances. (Evidently, it is not the dowsing rod itself that responds to the presence of water and other things, but the brain and nervous system of the person who holds the rod. The rod, pendulum, or other dowsing device does not move unless held by a dowser; they merely enlarge the subtle and involuntary muscle responses that move the dowser's arm.)

It appears that dowsers can also pick up information that is not produced by natural causes but is projected long-distance by the mind of another person. "Dowsable" lines, figures, and shapes can be created by the conscious intention of one person, and these lines, figures, and shapes affect the mind and body of distant persons who have not been told what has been created and where. Their rods move just as if the figures, lines, and shapes were due to natural causes immediately in front of them. This is the finding of a series of remote-dowsing experiments carried out in the past ten years by Jeffrey Keen, a renowned engineer, together with colleagues at the Dowsing Research Group of the British Society of Dowsers.

In a considerable number of experiments, the exact shapes created by the experimenter could be identified by the dowsers. It turned out that the shapes could be positioned with an accuracy of a few inches even when created thousands of miles away. The accuracy of positioning was not affected by the distance between the person creating the dowsable fields and the physical location of the fields: the same results were produced whether the experimenter created a dowsable shape a few inches or five thousand miles away. There was no difference whether the experimenter stood on the ground, was in an underground cave, flew in a plane, or was inside an electromagnetically shielded Faraday cage. Time did not seem to be a factor either: the fields were created faster than measurements could be taken, even over large distances. Time also proved irrelevant because the fields remained present and stable at all times after their creation. In one case they endured for more than three years. But they could be canceled if the person who created them wanted it.

Keen concluded that dowsable fields are created and maintained in an "Information Field that pervades the universe." The brain interacts with this field and perceives dowsable fields as holograms. This, according to Keen and the Dowsing Research Group, is an instance of nonlocal interaction between the brain and the field by different and even distant individuals.

Not only can people communicate with the minds of other people, but they can also interact with other people's *bodies*. Reliable evidence is becoming available that the conscious mind of one person can produce repeatable and measurable effects on the body (the "soma") of another. These effects, in turn, are known as *telesomatic*.

Telesomatic effects were known to traditional cultures: anthropologists call them "sympathetic magic." Shamans, witch doctors, and those who practice such magic (voodoo, for example) can act on the person they target, or they can act on an effigy of that person, such as a doll. The latter practice was widespread among traditional peoples. In his famous study *The Golden Bough,* Sir James Frazer noted that Native American shamans would draw the figure of a person in sand, ashes, or clay and then prick it with a sharp stick or do it some other injury. The corresponding injury was said to be inflicted on the person the figure represented. Observers found that the targeted person often fell ill, became lethargic, and sometimes even died.

There are positive variants of sympathetic magic today that are widely known and practiced. One variant is the kind of alternative medicine known as spiritual healing. The healer acts on the organism of his or her patient by "spiritual" means—that is, by sending a healing force or healing information. Healer and patient can be directly face to face, or miles apart; distance does not seem to affect the outcome. The effectiveness of this kind of healing may be surprising, but it is well documented. Renowned physician Larry Dossey calls the corresponding medical practice "Era III nonlocal medicine," suggesting that it is the successor to Era I biochemical medicine, and Era II psychosomatic medicine.

Another form of positively oriented sympathetic magic is healing by intercessory prayer. The effectiveness of prayer has been known to religious people and communities for hundreds and indeed thousands of years. But the credit for showing that it can be documented by controlled experiments is due to cardiologist Randolph Byrd. He undertook a ten-month computer-assisted study of the medical histories of patients at the coronary care unit at San Francisco General Hospital. As reported in the *Southern Medical Journal* in 1988, Byrd formed a group of experimenters made up of ordinary people whose only common characteristic was a habit of regular prayer in Catholic or Protestant congregations around the country. The selected people were asked to pray for the recovery of a group of 192 patients; another set of 210 patients, for whom nobody prayed, made up the control group. Neither the patients nor the nurses and doctors knew which patients belonged to which group. The people who were to pray were given the names of the patients and some information about their heart condition. As each person could pray for several patients, all patients had between five and seven people praying for them.

The results were significant. The prayed-for group was five times less likely than the control group to require antibiotics (three compared to sixteen patients); it was three times less likely to develop pulmonary edema (six versus eighteen patients); none in the prayed-for group required endotracheal intubation (while twelve patients in the control group did); and fewer patients died in the former than in the latter group (though this particular result was statistically not significant). It did not matter how close or far the patients were to those who prayed for them, nor did the manner of praying make any difference. Only the fact of concentrated and repeated prayer was a factor, without regard to whom the prayer was addressed and where the prayers took place. A subsequent experiment regarding the effect of remote prayer carried out under still more stringent conditions by a team of investigators headed by W. S. Harris showed similarly significant results.

Intercessory prayer and spiritual healing, together with other mind- and conscious intent-based practices yield impressive evidence regarding the effectiveness of telepathic and telesomatic information- and energy-transmission. The practices produce real and measurable effects, but classical medicine and the mainstream of Western science have no explanation for them.

FOUR

The Crucial Science Fable— In-formation in Nature

We now take up the task of identifying the answer to the puzzles of coherence encountered by scientists in the various domains of investigation. The answer, we shall see, is the presence in nature of the active and effective kind of information—"in-formation"—that links all things in the universe and creates quasi-instant connection among them. This is the "crucial science fable" that promises to solve the riddles of coherence and provide the foundation for a theory that is truly a theory of *everything*.

Our review of the puzzles encountered at the frontiers of science has set the stage for the quest to which Part One of this book is dedicated: to discover the basis for a scientifically founded I-TOE. We have gained an important insight. We have found that in order to account for a growing number of things and processes that are undoubtedly real and are likely to be fundamental, a new factor needs to be added to the repertory of laws and concepts of contemporary science. What is this new factor? Let us look at the principal findings:

- Astonishingly close connections exist on the level of the quantum: every particle that has ever occupied the same quantum state as another particle remains correlated with it in a mysterious, nonenergetic way.

- The universe as a whole manifests fine-tuned connections that defy commonsense explanation.
- Post-Darwinian evolutionary theory and quantum biology discover similarly puzzling connections within the organism, as well as between the organism and its milieu.
- The connections that come to light in the farther reaches of consciousness research are just as strange: they are connections between the consciousness of one person and the mind and body of another.

These connections indicate links between the particles that make up the material substance of the universe, as well as between the parts or elements of the integrated systems constituted of the particles. The links fine-tune the particles and the elements of the systems, creating space- and time-transcending coherence among them.

The surprising "nonlocal" forms of coherence crop up in fields as diverse as quantum physics, cosmology, evolutionary biology, and consciousness research. Some physicists—John Bell and Chris Clarke among them—suggest that nonlocality may be in fact the deeper reality; ordinary, so-called "classical" or "decoherent" states (states where things have a unique location and a unique set of physical characteristics) may appear merely as a consequence of the way we interact with medium-sized things—things that are neither as small as quanta nor as large as the cosmos.

Independently of the truth of such speculations, it is clear that nonlocal coherence has important implications. It signals that there is not only matter and energy in the universe, but also a more subtle yet real element: an element that connects, and which produces the observed quasi-instant forms of coherence.

Identifying this connecting element could solve the puzzles at the forefront of scientific research and point the way toward a more fertile paradigm. We can take the first step toward this goal by affirming that information is present, and has a decisive role, in all principal domains of nature. Of course, the information that is present in nature is not the everyday form of information but a special kind: it is "in-formation"—the active, physically effective variety that "forms" the recipient, whether it is a quantum, a galaxy, or a human being.

We explore this crucial "science fable" by identifying the physical origins of in-formation in nature. We join David Bohm, Harold Puthoff, and other scientists in looking for its roots in the complex and as yet not fully understood quasi-infinite energy sea called *quantum vacuum*.

THE QUANTUM VACUUM—OR PLENUM

The Dawn of the Physical Vacuum

Vacuum in its ordinary usage means "empty space." In cosmology it is used to refer to cosmic space in the absence of matter. In classical physics such space was considered passive, unsubstantial, and Euclidian, that is "flat." But in the nineteenth century physicists speculated that cosmic space is not truly empty: it is filled with the invisible energy field they called *luminiferous ether*. It was believed that the ether produces friction when bodies move through it and thus slows their motion. But this was a short-lived belief. At the turn of the twentieth century the famous Michelson-Morley experiments failed to observe the expected effect, and the ether was removed from the physicists' world picture. The absolute vacuum—space that is truly empty when not occupied by matter—took its place.

But the concept of space as empty did not dominate for long. Einstein's relativity theory wed space with time in a four-dimensional matrix that interacts with matter. Subsequent observations and experiments showed that this matrix has a physical reality of its own. In the "grand unified theories" (GUTs) developed in the second half of the twentieth century, the roots of all of nature's fields and forces are traced to the "unified vacuum." Thus the vacuum is neither empty space nor a purely geometrical structure: it is a physically real medium that interacts with matter and produces physically real effects.

This insight dawned gradually but inexorably in the course of the last forty years. In the 1960s Paul Dirac showed that fluctuations in fermion fields (fields of matter particles) produce a polarization of the zero-point field (ZPF) of the vacuum, whereby the vacuum in turn affects the particles' mass, charge, spin, or angular momentum. At around the same time, Andrei Sakharov proposed that relativistic phenomena (the slowing down of clocks and the shrinking of yardsticks near the speed of light) are the result of effects induced in the vacuum due to the shielding of the ZPF by charged particles. This was a revolutionary idea, since in this conception the vacuum is more than relativity theory's four-dimensional continuum: it is not just the geometry of space-time, but a real physical field producing real physical effects.

The physical interpretation of the vacuum in terms of the ZPF was reinforced in the 1970s, when Paul Davis and William Unruh put forward a hypothesis that differentiates between uniform and accelerated motion in the zero-point field. Uniform motion would not disturb the ZPF, leaving it isotropic (the same in all directions), whereas accelerated motion would produce a thermal radiation that breaks open the field's all-directional symmetry. During the 1990s, numerous explorations were undertaken on this premise.

Harold Puthoff, Bernhard Haisch, and collaborators produced a sophisticated theory according to which the inertial force, the gravitational force, and even mass are consequences of the interaction of charged particles with the ZPF. Puthoff claims that the very stability of atoms is due to interaction with the vacuum.

Mounting Evidence for the
Reality of the Physical Vacuum

Pressure waves have been found to propagate in interstellar space. Astronomers in NASA's Chandra X-ray Observatory found a wave generated by the supermassive black hole in the Perseus cluster of galaxies, some 250 million light-years from Earth. It has been traveling in the vacuum for the past 2.5 billion years. (The wave's frequency corresponds to the musical note B flat. Our ears cannot perceive it: it is fifty-seven octaves below middle C, more than a million billion times deeper than the limits of human hearing.) The fact that this wave is traveling in the vacuum is a further indication that the vacuum is neither empty nor passive: it is an active, physically real medium in which real-world events produce physically real waves.

Looking at phenomena at smaller scales, the physical reality of the vacuum remains equally evident. It turns out that life itself depends crucially on interactions with the vacuum. The evidence concerns the nature of the bonds among water molecules.

We know that living organisms consist of as much as 70 percent water. But it was not known that the properties of water make life itself possible. These properties do not derive from the chemical composition of the H_2O molecules of water; the decisive processes involve the bonds *between* the hydrogen components of the H_2O molecules. These bonds are more than ten times weaker than the typical chemical bonds. Because of the stretching of the molecular bonds between hydrogen atoms and their host oxygen atom, every drop of water is a constantly forming and re-forming mixture of molecular structures. Felix Franks of the University of Cambridge has shown that this flexibility is due to the interaction of the bonds with quantum-level vibrations in the zero-point field.

Additional support for the thesis that the vacuum is a complex and physically real medium is furnished by the current discussion on the Higgs field. The Higgs field (and the Higgs boson, the particle that is assumed to be associated with it) is different from all the other fields known to physics. In regard to all other fields, a region of space has the lowest possible energy when the energy of the field goes to zero. Not so in the case of the Higgs field. The lowest energy level of a region of space occurs when the energy of the Higgs field has a specific value that is *non*zero. This means that in the universe's lowest energy state its fields and forces are not at zero: in that "basic" and "most probable" state the universe is permeated by distinctly active fields.

Physicists postulate the nonzero Higgs field in order to explain one of the fundamental puzzles of their discipline. So-called matter particles have mass (even neutrinos, long thought to be massless, are now believed to possess some mass) but how they acquired their mass is by no means clear. The current hypothesis is that particles acquire mass through interaction with the Higgs field. The mass they acquire is proportionate to the strength of the Higgs field times the strength of their interaction. Even the mass of the mysterious dark matter in the universe is believed to arise through interaction with the Higgs field—more precisely, with a distinct variety of Higgs field. Without the different varieties of Higgs field there would be nothing for us to observe in the universe, nor would we be here to make the observations.

The Vacuum and the Fate of the Universe

The quantum vacuum turns out to be responsible for the fate of the universe as well. The universe could be *flat* (so that light—except near massive bodies—travels in a straight line), or *open* (with an infinitely expanding space-time that is negatively curved, like the surface of a saddle), or else *closed* (where expansion is overtaken by gravitation in a space-time that is positively curved like the surface of a balloon). In its future development it could continue to expand, or it could reverse, contract, and collapse, or else it could remain permanently balanced between expansion and contraction.

It was previously thought that the value of the gravitational force associated with massive particles ("matter") was the factor responsible for deciding which of these cosmic futures will come about. If there were more matter in the universe than the "critical density" (estimated at 5×10^{-26} g/cm^3), the gravitational pull associated with matter particles would exceed the inertial force generated by the Big Bang. Then the expansion of the galaxies would reverse, making this a closed universe. If, however, matter density were below the critical quantity, its gravitational pull would be more modest, and the force of expansion would continue to dominate it; we would then live in an open universe. But if matter-density were precisely at the critical value, the force of expansion would ultimately counteract the force of gravitation and thereafter the universe would remain precisely balanced at the razor's edge between expansion and contraction.

The question regarding matter-density, and hence whether the universe is open, closed, or flat, was to be decided in light of progressively more refined measurements. In 1998 the Boomerang (Balloon Observations of Millimetric Extragalactic Radiation and Geophysics) project's observations of the cosmic microwave background became available, followed by the observations of MAXIMA (Millimeter Anisotropy Experiment Imagining Array) and of DASI (Degree Angular Scale Interferometer, based on a microwave telescope at the South Pole). In February of 2003, the findings of the WMAP (Wilkinson Microwave Anisotropy Probe, a satellite launched in Earth orbit in June 30, 2001) were released. The same as the previous findings, they held no surprises, but refined the previous estimates and provided greater certainty of their validity. As a result, it now appears beyond reasonable doubt that we live in a flat universe.

However, the critical factor turned out to be not matter-density alone: the fate of the universe will be codetermined by the intrinsic energy of the vacuum. If the vacuum exerts a repulsive force, our flat universe will expand forever. On the other hand, if the energy of the vacuum is negative, the additional force of attraction will overcome the force of expansion and our universe will collapse.

When more precise information regarding the recession of distant galaxies became available, it turned out that not only is the universe expanding, it is expanding faster than cosmologists had thought. Clearly, the energy responsible for the expansion had to be factored into the equations. Cosmologists responded by reintroducing Einstein's "cosmological term" as the "cosmological constant." Their consensus is that the energies represented by the cosmological constant derive from the vacuum.*

There is no consensus, however, on the method to derive the proper value for the energies of the vacuum. The value derived from the equations of the Standard Model of particle physics is enormously greater than the energy required to account for the observed recession of the galaxies. If the energy of the vacuum were as great as the calculations indicate, so much energy would be injected in the universe that not only would distant galaxies recede, but all galaxies, and indeed all stars and planets, would instantly fly apart. The universe would expand like a rapidly inflating balloon. In our vicinity, space would be nearly empty.

The vacuum turns out to be a cosmic medium that transports photon-waves (light) as well as density-pressure waves, exerts the force that may ultimately decide the fate of the universe, and endows the particles we know as "matter" with mass. Such a medium is not an abstract theoretical entity. It is not a vacuum, but a physically real and active *plenum.*

---

*Einstein added a repulsive force factor to his general relativity equations in order to balance the contracting force of gravity, for he maintained that the universe is in a permanently balanced steady state (when it turned out that this is not the case, he gave up this tenet). Einstein called the expansive force "the cosmological term" and placed it on the left side of his general relativity equation, where it is associated with the space-time metric tensor—indicating that this force is a property of space. Currently the cosmological constant is placed on the right side of the general relativity equation where it is associated with the stress-energy tensor. This means that the cosmological constant is now treated as a manifestation of the energy of the vacuum.

"IN-FORMATION" IN THE QUANTUM VACUUM

The physical vacuum, effectively a cosmic plenum, subtends the observed universe. It transports light, energy, and pressure. Could it also be responsible for the remarkable coherence discovered at all scales in nature, from the quantic to the cosmic? Could it conserve and convey an active form of *information*?

This possibility has been raised by a number of avant-garde investigators. John Wheeler declared that information is even more fundamental in the universe than energy. Harold Puthoff wrote, ". . . on the cosmological scale a grand hand-in-glove equilibrium exists between the ever-agitated motion of matter on the quantum level and the surrounding zero-point energy field. One consequence of this is that we are literally, physically, 'in touch' with the rest of the cosmos as we share with remote parts of the universe fluctuating zero-point fields of even cosmological dimensions." And, he added, "Who is to say whether, for example, modulation of such fields might not carry meaningful information as in the popular concept of 'the Force'?"

The experiences of the Apollo astronaut Edgar Mitchell while in space led him to a similar conclusion. According to Mitchell, information is one part of a "dyad" of which the other part is energy. It is a part of the very substance of the universe. Information is present everywhere, he said, and has been present since the beginning. The quantum vacuum is the "holographic information mechanism that records the historical experience of matter." The information meant here is evidently active: it is "in-formation." The question is, just how does this in-formation mechanism in the quantum vacuum work: how does it record the "historical experience of matter"?

The answer to this seemingly mind-boggling question is surprisingly simple. We can spell it out in broad outline.

We know that interactions between things in the physical world are mediated by energy. Energy can take many forms—kinetic, thermal, gravitational, electric, magnetic, nuclear, and actual or potential—but in all its forms energy conveys some effect from one thing to another, from one place and one time to another place and another time. This is true, but it is not the whole truth. Energy must be conveyed by something; it does not act in a vacuum. Rather, scientists are now coming to the insight that it *does* act in a vacuum, namely in the *quantum* vacuum. That vacuum is far from empty: as we have seen, it is an active, physically real cosmic plenum. It conveys not only light, gravitation, and energy in its various forms, but also information; more exactly, "in-formation."

BACKGROUND BRIEF
WHAT IS "IN-FORMATION"?

What in-formation is *not*: The "theory of in-formation" is not the same as standard "information theory," because in-formation is not information in any of the scientific or everyday definitions. It is neither knowledge received about some fact or event, nor a pattern imposed on a transmission channel, nor yet the reduction of uncertainty regarding multiple choices. Information—in the sense of knowledge about things and events—may be conveyed by in-formation, but in-formation itself is different from information in the usual definitions.

What in-formation *is*: In-formation is a subtle, quasi-instant, non-evanescent and non-energetic connection between things at different locations in space and events at different points in time. Such connections are termed "nonlocal" in the natural sciences and "transpersonal" in consciousness research. In-formation links things (particles, atoms, molecules, organisms, ecologies, solar systems,

---

entire galaxies, as well as the mind and consciousness associated with some of these things) regardless of how far they are from each other and how much time has passed since connections were created between them.

In-formation conveyed in and through the vacuum can account for the puzzling forms of coherence we find in the various domains of nature. How this process takes place can be reconstructed on the basis of theories advanced at the cutting edge of the new physics.

A much discussed theory advanced by the Russian physicists G. I. Shipov, A. E. Akimov, and colleagues gives a mathematically elaborated account of the linking of proximal or distant events by the "physical vacuum." The gist of their theory is that charged particles "excite" the ground state of the vacuum and create minute vortices in it. The resulting field is a system of rotating wave packets of electrons and positrons. Where the wave packets are mutually embedded, this "torsion field" is electrically neutral. If the embedded packets have opposite spins, the system is compensated not only in charge, but also in classical spin and magnetic moment. Such a system is a "phyton." Dense ensembles of phytons approximate a model of the vacuum's torsion field. The vortices of this field carry information, linking particles at the mind-boggling group-speed of 10^9 c, that is, one billion times the speed of light.

The theory advanced by the Hungarian theoretician Laszlo Gazdag provides an analogous explanation. It takes as its basis the well-known fact that particles that have the quantum property known as "spin" also have a magnetic effect: they possess a specific magnetic momentum. The magnetic impulse, Gazdag suggests, is registered in the vacuum in the form of minute vortices. Like vortices in water, vacuum-based vortices have a nucleus around which circle other elements—H_2O molecules in the case of water, virtual bosons (force particles) in the case of the zero-point field. These tiny vortices carry information, much as magnetic impulses do on a computer disk. The information carried by a given vortex corresponds to the magnetic momentum of the particle that created it: it is information on the state of that particle.

These minute spinning structures travel in the vacuum, and they interact with each other. When two or more vortices meet, they form an interference pattern that integrates the strands of information on the particles that created them. This interference pattern carries information on the entire ensemble of the particles that produced the vortices.

THE PARABLE OF THE SEA

The above are abstruse theories, but their message can be conveyed in everyday terms. Let us take the example of the sea. When a ship travels on the sea's surface, waves spread in its wake. These affect the motion of all other ships in that part of the sea. Every ship—and every fish, whale, or object in that part of the sea—is exposed to these waves and its path is in a sense "informed" by them. All vessels and objects "make waves," and their wave fronts intersect and create interference patterns.

If many things move simultaneously in a waving medium, that medium becomes modulated: full of waves that intersect and interfere. This is what happens when several ships ply the sea's surface. When we view the sea from a height—a coastal hill or an airplane— we can see the traces of all the ships that passed over that stretch of water. We can also see how the traces intersect and create complex patterns. The modulation of the sea's surface by the ships that ply it carries information on the ships themselves. It is possible to deduce the location, speed, and even the tonnage of the vessels by analyzing the interference patterns of the waves they have created.

As fresh waves superimpose on those already present, the sea becomes more and more modulated; it carries more and more information. On calm days we can see that it remains modulated for hours, and sometimes for days. The wave patterns that persist are the memory of the ships that traveled on that stretch of water. If wind, gravity, and shorelines did not cancel these patterns, the wave-memory of the sea would persist indefinitely.

The theories of Shipov, Akimov, and Gazdag give a scientific formulation of the process of wave-formation and wave-memory creation in a medium that is not the ordinary sea, but the extraordinary quantum vacuum. The vortices generated in the vacuum propagate as torsion wavefields. The wavefields meet and create wave-interference patterns. These contain information on the state of the particles that created the vortices—and their joint interference pattern holds information on the ensemble of the particles whose torsion-wavefields have interfered. In this way the vacuum carries information on atoms, molecules, macromolecules, cells, and even on organisms and populations and ecologies of organisms. There is no evident limit to the information that interfering vacuum wavefields could conserve. In the final count, they could carry information on the state of the whole universe.

We should note that the information carried in the vacuum is not localized, confined to a single location only. As in a hologram, the vacuum carries information in distributed form, present at all points where the wavefields have propagated. The interfering wavefields in the vacuum are natural holograms. They propagate quasi-instantly, and nothing can attenuate or cancel them. Thus nature's holograms are cosmic holograms: they link—"in-form"—all things with all other things.

A BRIEF ON HOLOGRAMS

Holograms are three-dimensional representations of objects recorded with a special technique. A holographic recording consists of the pattern of interference created by two beams of light; monochromatic lasers and semitransparent mirrors function best for this purpose. Part of the laser light passes through the mirror and part is reflected and bounced off the object to be recorded. A photographic plate is exposed with the interference pattern created by the light beams. This is a two-dimensional pattern and it is not meaningful in itself; it is merely a jumble of lines. Nonetheless, it contains information on the contours of the object. These contours can be re-created by illuminating the plate with laser light. The patterns recorded on the photographic plate reproduce the interference pattern of the light beams, so that a visual effect appears that is identical to the 3-D image of the object. This image appears to float above and beyond the photographic plate, and it shifts according to the angle at which one views it.

Interestingly and importantly, the image appears no matter what part of the holographic plate is illuminated, though it becomes less distinct as the illuminated area is reduced. The fact is that the information on which the image is based is present throughout the holographic record.

ENTER THE AKASHIC FIELD

The idea that information is present throughout nature is a recurrent theme in cultural history, but it is new to Western science. It calls for the recognition that information is not an abstract concept: as "in-formation" it has a reality of its own. It is a part of the physical universe. And because it is present throughout nature, it is best conceptualized as an extended *field*.

Rationale for an In-formation Field

The evidence for a field that would conserve and convey information is not direct; it must be reconstructed in reference to more immediately available evidence. Like other fields known to modern physics, such as the gravitational field, the electromagnetic field, the quantum fields, and the Higgs field, the in-formation field cannot be seen heard, touched, tasted, or smelled. However, this field produces effects, and these can be perceived. This is the same in regard to all the fields known to science. For example, the gravitational or G-field cannot be perceived: when we drop an object to the ground, we see the object falling but not the field that makes it fall—we see the effect of the G-field but not the G-field itself. The effect of the G-field is gravitation among separate masses; general relativity and related field theories seek to show that the G-field is the simplest and most consistent explanation of the effects. The same applies to the electromagnetic or EM-field, of which the effect is the transmission of electric and magnetic force; to the Higgs field, of which the presumed effect is the presence of mass in particles; and to the weak and strong nuclear fields, where the effect is attraction and repulsion among particles at extreme proximity to each other.

In the case of the field that could account for the presence of information in nature, the evidence is the puzzling, quasi-instant form of coherence that comes to light in the physical, cosmological, and biological sciences, as well as in consciousness research. These phenomena call for an explanation, and the simplest and most logical explanation is a field that links the entities that prove to be nonlocally coherent.

The concept of an in-formation field is new to science but it is concordant with its history. In the history of modern science, the idea that things and events could be affecting one another without being connected by some physically real medium has been rejected. Entities that turned out to be linked with one another across space (and perhaps also over time) have been said to be connected by an intervening physical field. Michael Faraday, for example, proposed that electric and magnetic phenomena are linked by an electric and a magnetic field—and that this is one and the same field: the electromagnetic field.

Faraday's electromagnetic field was seen as a local field, associated with the given objects. James Clerk Maxwell proposed that the electromagnetic field is not local but universal: it is present everywhere. Modifications of the EM-field travel throughout space at the speed of light. A changing electric field produces changes in the magnetic field, and this in turn produces changes in the electric field.

The universal electromagnetic field was a revolutionary insight, for it meant abandoning the notion of empty space as a mere vehicle for conveying the forces involved in the interaction of particles. Space was henceforth conceived as a continuous universal field through which electric and magnetic effects are transmitted between particles, whether they are contiguous in space or removed from each other.

The explanation of the mutual attraction of massive objects has a similar history. In Newton's theory gravitation is a local phenomenon, an intrinsic property of objects with mass (though Newton was greatly puzzled by this property, as was Ernst Mach subsequently). In his general theory of relativity Einstein removed the gravitational force from individual objects and ascribed it to space-time itself: gravitation was henceforth considered a universal field.

As we have seen, another universal field has recently entered the world-picture of physics: the Higgs field. For the present, the Higgs field is deduced from the mathematical structure of particles and of particle interactions as given in the Standard Model of particle physics (although experimental evidence is expected to become available when accelerators powerful enough to reach the estimated energy level of the "Higgs boson" come on line). Similarly to gravitation, the Higgs field, too, has to do with mass, but not with the property of massive objects: this field is to account for the very *existence* of mass.

The history of the field concept demonstrates that when phenomena occur that require a physical explanation, scientists first attempt to give an explanation specifically related to the entities that manifest the phenomena. As theories grow and develop, the explanatory concepts tend to become more general. In this way, what were initially seen as local force fields are later understood as universal fields, present at all points in space and time. Electric and magnetic phenomena are now ascribed to the universal EM-field; the mutual attraction of noncontiguous objects is ascribed to the universal G-field; and the presence of mass is ascribed to the universal Higgs field.

The time has come to add another field to science's repertory of universal fields. Although fields, like other entities, are not to be multiplied beyond the scope of necessity, it seems evident that a further field is required to account for the special kind of coherence revealed at all scales and domains of nature, from the microdomain of quanta, through the mesodomain of life, to the macrodomain of the cosmos. This field is not the zero-point field, for its properties transcend those currently thought to be associated with that field. It is a different field, of which we know the effects but do not yet possess a mathematical description. It is nonetheless clear that this field exists, for it produces real effects. Just as electric and magnetic effects are conveyed by the EM-field, attraction among massive objects by the G-field, and attraction and repulsion among the particles of the nucleus by quantum fields, so we must recognize that a universal in-formation field conveys the effect we described as "nonlocal coherence" throughout the many domains of nature.

The Akashic Field

In his previous books this writer named the in-formation field the *Akashic Field,* or *A-field* for short. What is the reason for this name?

In the Sanskrit and Indian cultures, Akasha is an all-encompassing medium that *underlies* all things and *becomes* all things. It is real, but so subtle that it cannot be perceived until it becomes the many things that populate the manifest world. Our bodily senses do not register Akasha, but we can reach it through spiritual practice. The ancient Rishis reached it through a disciplined, spiritual way of life, and through yoga. They described their experience and made Akasha an essential element of the philosophy and mythology of India.

In the twentieth century Akasha has been brilliantly described by the great Indian Yogi Swami Vivekananda.

> According to the philosophers of India, the whole universe is composed of two materials, one of which they call Akasha. It is the omnipresent, all-penetrating existence. Everything that has form, everything that is the result of combination, is evolved out of this Akasha. It is the Akasha that becomes the air, that becomes the liquids, that becomes the solids; it is the Akasha that becomes the Sun, the Earth, the Moon, the stars, the comets; it is the Akasha that becomes the human body, the animal body, the plants, every form that we see, everything that can be sensed, everything that exists. It cannot be perceived; it is so subtle that it is beyond all ordinary perception; it can only be seen when it has become gross, has taken form. At the beginning of creation there is only this Akasha. At the end of the cycle the solid, the liquids, and the gases all melt into the Akasha again, and the next creation similarly proceeds out of this Akasha. . . .

The sum total of all forces in the universe, mental or physical, when resolved back to their original state, is called Prana. When there was neither aught nor naught, when darkness was covering darkness, what existed then? Then Akasha existed without motion. . . . At the end of a cycle the energies now displayed in the universe quieted down and became potential. At the beginning of the next cycle they start up, strike upon the Akasha, and out of the Akasha evolve these various forms . . .

The reason for naming the in-formation field in nature the Akashic field should now be evident. The Akashic vision of a cyclic universe—of a Metaverse that creates universe after universe—is essentially the vision we now get from cosmology. In the new physics the unified, physically real vacuum is the equivalent of Akasha. It is the original field out of which emerged particles and atoms, stars and planets, human and animal bodies, and all the things that can be seen and touched. It is a dynamic, energy-filled medium in ceaseless fluctuation. The vacuum is Akasha and Prana rolled into one—the womb of all the "matter" and all the "force" in the universe.

The maverick genius Nikola Tesla adopted this vision in the context of modern science. He spoke of an "original medium" that fills space and compared it to Akasha, the light-carrying ether. In his unpublished 1907 paper "Man's greatest achievement," he wrote that this original medium, a kind of force field, becomes matter when Prana, cosmic energy, acts on it, and when the action ceases, matter vanishes and returns to Akasha. Since this medium fills all of space, everything that takes place in space can be referred to it.

For Tesla the idea of curved space—put forward at the time by Einstein—was not the answer. However, most physicists adopted Einstein's mathematically elaborated four-dimensional curved space-time and refused to consider the concept of a space-filling medium or force field. Tesla's insight fell into oblivion. Today, a hundred years later, it has been revived. Scientists now realize that space is not empty, and what is called the quantum *vacuum* is in fact a cosmic *plenum*. It is a fundamental medium that recalls the ancient concept of Akasha.

In the next development of science, the A-field will join the currently known universal fields: the G-field, the EM-field, the Higgs field, and the locally effective but universally present strong and weak nuclear fields.

PART TWO

THE IN-FORMED UNIVERSE

Perennial Questions and Fresh Answers from the Integral Theory of Everything

Beyond the puzzle-filled world of the mainstream sciences, a new concept of the universe is emerging. The established concept is transcended; in its place comes the in-formed universe, rooted in the rediscovery of ancient tradition's Akashic Field as the vacuum-based holofield.

In this concept the universe is a highly integrated, coherent system: a "supermacroscopic quantum system." Its crucial feature is in-formation that is generated, conserved, and conveyed, and links all its parts. This feature is entirely decisive. It transforms a universe that is blindly groping its way from one phase of its evolution to the next into a strongly interconnected system that builds on the in-formation it has already generated.

In the in-formed universe the A-field is a fundamental element. Thanks to in-formation conserved and conveyed by the A-field, the universe is of mind-boggling coherence. All that happens in one place happens also in other places; all that happened at one time happens also at all times after that. Nothing is "local," limited to where and when it is happening. All things are global, indeed cosmic, for all things are connected, and the memory of all things extends to all places and to all times.

This is the concept of the in-formed universe, the view of the world that will hallmark science and society in the coming decades.

FIVE

The Origins and Destiny
of Life and the Universe

In Part Two we query the integral theory of everything we developed in Part One. We ask some of the "great questions" thinking people have always asked about the world in which we live and explore the answers we get in the in-formed universe. In this chapter we ask: Where did everything come from? Where is it going? Is there life elsewhere in the wide reaches of the universe? If so, is it likely to evolve to higher stages or dimensions?

WHERE EVERYTHING CAME FROM—
AND WHERE IT IS GOING

Perhaps the most fundamental question ever asked is, *Where did the universe come from?*

The earliest answers were couched in the mystical worldview, followed by the worldviews of the great religions. In regard to concepts of origin and destiny, the early intuitions of East and West were remarkably consistent: they both envisaged the origins of the universe as a stupendous process of self-creation. But with the rise of monotheistic religion in the West, the creation story of the Old Testament replaced mystical and metaphysical accounts. Throughout the Middle Ages, Christians, Muslims, and Jews believed that an all-powerful

God created the sky above and the Earth below, and all things in between, with purpose and intent, just the way we find them.

In the nineteenth century, the Judeo-Christian account of creation came into conflict with the theories of modern science, in particular with Darwinian biology. A vivid contrast arose between the view that everything we behold was created intentionally by a divine power and the concept according to which living species evolve on their own, from simpler common origins. The contrast fueled endless debates, surviving to this day in the controversy surrounding the teaching of "creationist" versus "evolutionist" theories in public schools.

Since the 1930s, the Judeo-Christian creation story has had to contend not only with the Darwinian doctrine of biological evolution, but also with physical cosmology. Newton's clockwork universe required a Prime Mover to wind it up and get it going, and this could be attributed to the work of a Creator. Subsequently Einstein's steady-state universe could do without a Creator, for it persisted from the beginnings of time the same as it is today. But when the steady-state universe was replaced by the Big Bang theory's explosively expanding universe, questions arose again about the world's origins. If the universe was born in a Big Bang 13.7 (or, as a new finding indicates, 15.8) billion years ago and will end either in the Big Crunch some two thousand billion years in the future or in the evaporation of the last galactic-cluster-sized black holes at the almost inconceivable time horizon of 10^{122} years, the question that comes to mind is: *What was there before all this began—and what will be there after it is over?*

ORIGINS AND EVOLUTION
OF OUR UNIVERSE

The standard cosmology, known as BB theory, can say no more about how the universe came into being than that a random instability took place in a fluctuating cosmic vacuum, the pre-space of the universe. It cannot say either why this instability occurred or why it occurred *when* it occurred. And otherwise than through implausibly speculative fables—such as a cosmic roulette among a large number of randomly created universes—it also cannot say *why* our universe came to be the way it came to be: why it has the remarkable properties it now exhibits. The question returns, it seems, to the domain of religion and mysticism. But giving up on science would be premature. The Big Bang theory is not the final word; the new cosmologies have more to say about cosmic origins.

As we have seen, there are sophisticated cosmologies that tell us that our universe is not the only universe. There is also a meta-universe or Metaverse that was not created in the Bang that created our universe (which was but one of many explosions, so it no longer qualifies for the adjective "Big"); nor will the Metaverse itself come to an end when the particles created by this particular Bang vanish in the collapse of the last black holes. The insight that dawns is that *the* universe existed prior to the birth of *our* universe, and it will continue to exist after our universe's demise. *The* universe is the Metaverse, the mother of our universe and perhaps of myriad other universes.

Cosmologies of the Metaverse are in a better position than the Big Bang theory (which is limited to our universe) to speak of conditions that reigned before, and will reign after, the life cycle of our universe. The quantum vacuum, the subtle energy and in-formation sea that underlies all "matter" in the universe, did not originate with the Bang that produced our universe, and it will not vanish when the particles created by that explosion fall back into it. The subtle energies and the active in-formation that underlie this universe were there before its particles appeared and will be there after they disappear. The deeper reality is the quantum vacuum, the enduring in-formation and energy sea that pulsates, producing periodic explosions that give rise to local universes.

Universe-creating explosions (recurring "Bangs") are instabilities in the Metaverse's vacuum. The Bangs create pairs of particles and antiparticles, and the surviving surplus of particles populates the newborn universe's space-time. The particles cohere into atoms, and in time gravitation clumps together the particles and atoms in galactic and stellar structures, and the kind of evolution we observe in our universe gets under way. It unfolds time after time.

The actual evolution of universes is conditioned by the interplay of the gravitational attraction among massive particles and the repulsive or attractive energies of the vacuum itself. We have no certainty regarding the exact outcome of this interplay in our own universe, and in any case other universes could be born with different parameters and hence with different outcomes. Yet whether the evolution of particular universes results in continual expansion, expansion followed by contraction, or a balance between the forces of expansion and contraction, the end of "matter" in a universe remains the same. Following the exhaustion of their nuclear fuel, stars either explode or collapse. Ultimately, later generations of stars collapse and become quasars and black holes. Galaxies themselves collapse on themselves as black holes form at their center, such as the recently discovered black hole at the center of our Milky Way galaxy. Sooner or later all galaxies "evaporate" in supergalactic black holes, with the degenerate remnants of their atoms vanishing into the vacuum.

The explosive exit of matter in supergalactic black holes could be the prelude to matter-creating explosions. "Star-bursts" of this kind have been observed and some of them could produce enough matter to become universes of their own.

Notwithstanding technical disagreements among different cosmological scenarios, most cosmologists agree that we live in a cyclically creative/destructive multiverse rather than in a single-cycle universe. Local universes evolve, die back, and coexist with, or are succeeded by, other universes in the embrace of a vast, temporally (if not necessarily spatially) infinite universe that endures over the entire cycle: the Metaverse.

Some cosmologies assume that local universes are isolated from each other. Yet if the universes had no causal contact with one another, each of them would start with an accidental configuration of its basic laws and constants. Such a randomly configured universe has negligible chances of giving rise to complex systems such as living organisms. If we assume that at its birth our universe was fully isolated from other universes we cannot find a natural explanation for its astonishing propensity to bring forth life. Scientists could only marvel at the incredible serendipity that life could arise and evolve on Earth, and hand over the question to poets and prophets.

Instead of marveling at this improbable scenario and giving up on a scientific explanation of it, we can contemplate the possibility that at its birth our universe was in-formed by a universe or universes that existed prior to it. This is not an unscientific assumption. All universes that may exist, and have ever existed, must have arisen from the cosmic vacuum. The universe or universes that preceded ours "excited" the vacuum and created wave-interference-based holograms in it. These then affected—"in-formed"—the evolution that took place in the succeeding universes. The systems that evolved in those universes further affected—in-formed—the vacuum. Thus, through the vacuum an ongoing transfer of in-formation came about between the universes. In the cycle of universes in the Metaverse each universe is in-formed by its predecessor, and in-forms in turn its successor.

The in-formation transmitted from a precursor to a successor universe affects the value of the energy of the vacuum and determines the amount of matter in the successor universe. It also affects the distribution of the virtual states that particles, atoms, molecules, and systems and assemblies of molecules can occupy when they leap from "virtual" to "real" states. This distribution determines in turn the kind of interactions that the particles and systems of particles can enter into, and hence the kind of systems that can result from the interactions. In this way each universe "inherits" the physical properties of its precursor. It neither collapses back on itself shortly after its birth nor expands so fast that only a dilute gas of particles survives. It evolves more and more efficiently, and hence further and further into the otherwise improbable realms of coherence where complex systems such as organisms, societies, and ecologies can emerge.

Thus at its origins our universe did not come by the fine-tuned properties we observe by chance: it "inherited" them from a prior universe. What about the properties of the Metaverse itself? Can we explain properties that not only give rise to a coherently evolving universe, but to an entire series of sequentially, and always more coherently, evolving universes?

In considering this awesome question we should start with what we already know and apply it to what we do not know and cannot know—at least, not in direct reference to what we experience. What we do know is that complex systems are "initial-condition dependent"—that is, their functioning and further development is strongly influenced by the circumstances under which they have come into existence. Our universe is a complex system, and its development must have been critically influenced by the conditions under which it was initiated—that is, by the "in-formation" of the vacuum in which it was born. This was the factor that fine-tuned the physical constants of our universe, and set the values of the laws of interaction that had led to our universe's micro- and macro-structures—its particles, atoms, and molecules, and its stars and galaxies.

The Metaverse's multicyclic evolution must have been critically influenced by its own initial conditions. Yet prior universes could not have set these conditions, for the Metaverse was there before all universes—its vacuum was primeval, virginal. How, then, were the initial conditions of the Metaverse created—*by what* . . . or is the question *by Whom*? This is the deepest and greatest mystery of all— the mystery of the origins of the universe-generating process itself.

This greatest of all mysteries is "transempirical"; it is not amenable to resolution by reasoning based on observation and experiment. Yet one thing is clear: If it is unlikely that our fine-tuned universe would have originated in a randomly configured vacuum, the mother universe that gave rise to a series of progressively evolving local universes is even more unlikely to have originated in a random, non-informed state.

The vacuum of the Metaverse was not only such that one universe could arise in it, but such that an entire series of universes could. This could hardly have been a lucky fluke. In some way, the primordial vacuum must have been already in-formed. *There must have been an original creative act, an act of "metaversal Design."*

DESIGN OR EVOLUTION?

THE CREATIONIST CONTROVERSY IN A NEW LIGHT

The persistent debate among conservative Christians, Muslims, and Jews (the "creationists") and scientists and the science-minded public (the "evolutionists") centers on biological evolution. But on a deeper look, it concerns the universe itself in which life evolved—or in which it was created.

At first glance, the science community—and anyone believing that science discloses some basic truth about the nature of reality—is compelled to reject the hypothesis that living species are the way they are because they were designed to be that way . . . that they are the result of special acts of creation. Yet it is also evident that it is highly unlikely that living species could have come about through processes of random mutation and natural selection. Affirming this theory, the creationists claim, makes the entire doctrine of evolution misguided.

Mainline Darwinists expose themselves to the objection of the creationists by contending that random processes of evolution are adequate to explain the facts. Richard Dawkins, for example, claims that the living world is the result of processes of piecemeal trial and error without deeper meaning and significance. Like Weinberg, Dawkins claims that there is no purpose and meaning to this world. Therefore, there is no need to assume that it was purposefully designed.

Take cheetahs, he said. They give every indication of being superbly designed to kill antelopes. The teeth, claws, eyes, nose, leg muscles, backbone, and brain of a cheetah are all precisely what we should expect if God's purpose in creating cheetahs was to maximize deaths among antelopes. At the same time, antelopes are fast, agile, and watchful, apparently designed so they can escape cheetahs. Yet neither the one nor the other feature implies creation by special design: Dawkins tells us that this is just the way nature is. Cheetahs have a "utility function" to kill antelopes, and antelopes, to escape cheetahs. Nature itself is indifferent to their fate. Ours is a world of blind physical forces and genetic replication where some get hurt and others flourish. It has precisely the properties we would expect it to have if at bottom there was no design, no purpose, and no evil and no good, only blind and pitiless indifference.

Evidently, if this were the case, it would be hard to believe in an intelligent Creator. The God that created the world would have to be an indifferent God, if not actually a sadist who enjoys blood sports. It is more reasonable, according to Dawkins, to hold that the world just is, without reason and purpose. The way it is results from random processes played out within limits set by fundamental physical laws. The idea of design is superfluous. In this regard, Darwinists echo the French mathematician Pierre Laplace, who is reputed to have told Napoleon that God is a hypothesis for which there is no longer any need.

Creationists point out, however, that it is entirely improbable that all we see in this world, ourselves included, should be the result of random processes governed by impersonal laws. The tenet that everything evolved by blind chance out of common and simple origins is mere theory, they say, unsubstantiated by solid evidence. Scientists cannot come up with manifest proof for this theory of evolution: "You can't go into the laboratory or the field and make the first fish," said Tom Willis, director of the Creation Science Association for Mid-America. The world around us is far more than a chance concatenation of disjoined elements; it exhibits meaning and purpose. This implies design.

The creationist position would be the logical choice if cutting-edge science asserted that the evolution of living species is the product of blind chance. But cutting-edge science does not assert this. As we have seen, post-Darwinian biology has discovered that biological evolution is not merely the outcome of chance mutations exposed to natural selection. The coevolution of all things with all other things in the planet's web of life is a systemic process with an orderly, nonrandom dynamic. It is part of the evolution of the universe from particles to galaxies and stars with planets. On Earth this evolution produced physical, chemical, and thermal conditions that were just right for the stupendous processes of biological evolution to take off. Such conditions could have come about only in a universe governed by precisely coordinated laws and regularities. Even a minute difference in these laws and constants would have foreclosed the emergence of life forever.

Thus, the debate between creationists and evolutionists shifts from the question regarding the origins of *life* to the question concerning the origins of the *universe*. In the last analysis, it shifts to the origins of the *Metaverse* in which our universe arose. Could it be that the Metaverse, the mother of our universe and of all universes past, present, and future, has been purposefully designed so that it could produce universes that give rise to life? For creationists, this is the simplest and most logical assumption. Evolutionists cannot object: evolution, being an irreversible process, must have had a beginning, and that beginning must be accounted for. It could not have been something out of nothing—a "free lunch"!

In the final count, the evolutionist/creationist controversy has no point. The question "Design or evolution?" poses a false alternative. Design and evolution do not exclude each other; indeed, they require one another. The Metaverse is unlikely to have come into existence out of nothing, as a result of pure chance. And if it—more exactly, its primordial vacuum—was already "in-formed," the Metaverse was in a sense *designed* to give rise to a series of sequentially evolving universes.

The bottom line is not "design *or* evolution." It is "design *for* evolution."

Where is the universe going? We now reverse the direction of our inquiry. Instead of moving back in time, we move forward. In a coherent, nonrandomly evolving universe this, too, is possible. The question we ask is: *Where is the evolution of this universe, and of all universes in the Metaverse, leading—to what ultimate state or condition?*

In contemplating this question we should remember that we are querying destiny and not fate. There is a fundamental difference between a point of origin and a point of destiny. A point of origin is in the past, and must be assumed to have been a definite and unique state. A point of destiny will likewise be a definite and unique state when it is reached—but it will not be that until it *is* reached. Much like the multipotentiality of the quantum that is free to choose its real state from among its virtual states until an interaction collapses its wave function, the cosmos will not have a determinate final state until it actually *reaches* that state. Not being classically mechanistic, it is undetermined as regards the choice of its ultimate state.

The cosmos has various possibilities for its evolution. The past is a stubborn fact, established once and for all, but the future is not. There is no certainty even regarding the ultimate fate of the universe: whether it will expand forever, collapse on itself, or remain balanced between expansion and contraction. But even if the evolution of the universe is uncertain, evolution *within* the universe can have an overall direction. This is because this universe is coherent and consistent: in it one thing entails another. When one choice is made, the cascade of consequences continues until the final state is reached. There is no need for setting a specific goal at the beginning: the goal is generated in the process itself. It is one toward which evolution in this universe generally tends; it is what gives it overall direction. This direction is toward greater and greater coherence and complexity.

A GAME THAT GENERATES
ITS OWN GOAL

The variant of the popular parlor game "twenty questions" suggested by John Wheeler (though he had an abstruse problem of quantum physics in mind) illustrates a process that heads toward a specific goal even though that goal was not given at the beginning.

In the usual version of this game, a person leaves the room and the others decide on a thing or object that the person is to guess. The latter can ask a maximum of twenty questions, and only "yes" or "no" answers can be given to each question. But each question narrows the scope of possibilities because it excludes alternative possibilities. For example, if the first question is "Is it living?" (as opposed to nonliving), a yes answer excludes all things other than plants, animals, insects, and simple organisms.

In the alternative version, a person leaves the room and the others, without telling him, agree not to agree on a given thing or object but pretend that they did. They must give consistent answers, however. Consequently, when the innocent interlocutor returns and asks, "Is it living?" and if the answer he or she gets is yes, then all subsequent answers must pretend that the thing to be guessed is a plant, an animal, or perhaps a microorganism. A skilled player can narrow the scope of possibilities in such a way that within twenty questions he or she identifies one definite answer— for example, the kitten next door. Yet that was not the goal when the game was started. There was no goal—the one that emerged was generated by the game itself!

The coherence and complexity-directed evolutionary process is not likely to be unique to *our* universe. It is highly improbable that our universe—which is so fine-tuned for the evolution of complexity—was the first universe to arise in the Metaverse. And if it was not the first universe to be born in the Metaverse, it is not likely to be the last. Other universes will arise in time. How will the process of evolution unfold across this stupendous cycle of universe after universe? We can extrapolate a general answer to this question as well.

We begin by noting that the evolution of universes within the Metaverse is cyclic but not repetitive. One universe informs another; there is progress from universe to universe. Each universe is more evolved than the one before. The cycle itself evolves from a random initial universe, to universes where the physical parameters are more and more tuned to the evolution of complexity. Thus cosmic evolution is toward universes where complex and coherent structures emerge, including structures that harbor evolved forms of life—and the evolved forms of mind that are presumably always associated with evolved forms of life.

The cycle of universes in the Metaverse progresses from universes that are purely *physical* to universes that include life. These are *physical-biological* universes. And given that forms of mind are associated with forms of life, the cycle leads from physical to physical-biological to *physical-biological-psychological* worlds.

Is reaching a physical-biological-psychological universe the deeper meaning of the evolution of the cycle of universes—of the Metaverse itself? Possibly, and even probably. But a definitive answer is foreclosed to science, and to any reasoning this side of mystical intuition.

LIFE ON EARTH AND IN THE UNIVERSE

We move now to the next set of "great questions": questions that are still "great" but somewhat more modest. They are questions about the origins and destiny of life on Earth and in the cosmos. The first query concerns the prevalence of life. *Is there life elsewhere in the universe—or is it unique to this planet?*

We have every reason to believe that the kind of life we know on Earth is not limited to this planet. Life arose here over four billion years ago, and since then it has been evolving inexorably, if highly discontinuously, building structure upon structure, system within and with system. We have no reason to doubt that wherever suitable conditions are present, processes of physical, physical-chemical, and ultimately biological and ecological self-organization are getting under way. And we have every reason to believe that suitable conditions have been and are present in many places. Astronomical spectral analysis reveals a remarkable uniformity in the composition of matter in stars and hence in the planets that are associated with stars. The most abundant elements are, in order of rank: hydrogen, helium, oxygen, nitrogen, and carbon. Of these, hydrogen, oxygen, nitrogen, and carbon are fundamental constituents of life. Where these occur in the right distribution and energy is available to start chains of reaction, complex compounds result. On many planets the active star with which the planet is associated furnishes such energy. The energy is in the form of ultraviolet light, together with electric discharges, ionizing radiation, and heat.

About four billion years ago, photochemical reactions took place in the upper regions of the young Earth's atmosphere, and the reaction products were transferred by convection to the surface of the planet. Electric discharges close to the surface deposited the products in the primeval oceans, where volcanic hot springs supplied further energy. The combination of energy from the Sun with energy stored below the surface catalyzed a series of reactions of which the end products were organic compounds. With local variations, the same system-building process is no doubt unfolding on other planets. Numerous experiments pioneered by the paleobiologist Cyril Ponnamperuma and others show that when conditions similar to those that were present on the primeval Earth are simulated in the laboratory, the very compounds emerge that form the basis of earthly life.

There must be other planets with conditions similar to those on Earth. There are more than 10^{20} stars in our universe, and during their active phase they all generate energy. When these energies reach the planets associated with the stars, they are capable of fueling the photochemical reactions required for life. Of course, not all stars are in the active phase, and not all have planets with the right chemical composition, of the right size, and at the right distance.

Just how many potentially life-bearing planets are there? The estimates vary. Taking a conservative tack, the Harvard astronomer Harlow Shapley assumed that only one star in a thousand has planets and that only one of a thousand of these stars has a planet at the right distance from it (in our solar system, there are two such planets). He further supposed that only one out of a thousand planets at the right distance is large enough to hold an atmosphere (in our system, seven planets are large enough), and that only one in a thousand planets at the right distance and of the right size has the right chemical composition to support life. Even then there should be at least 100 million planets capable of supporting life in the cosmos.

The astronomer Su-Shu Huang made less limiting assumptions and reached an even more optimistic estimate. He took the time scales of stellar and biological evolution, the habitable zones of planets and related dynamic factors, and came to the conclusion that no less than 5 percent of all solar systems in the universe should be able to support life. This means not 100 million, but 100 *billion* life-bearing planets. Harrison Brown came up with a bigger number still. He investigated the possibility that many planetlike objects that are not visible exist in the neighborhood of visible stars—perhaps as many as sixty such objects more massive than Mars. In that case, almost every visible star possesses a partially or wholly invisible planetary system. Brown estimated that there are at least 100 billion planetary systems in our own galaxy alone—and there are 100 billion galaxies in this universe. If he is right, life in the cosmos is immensely more prevalent than has been previously estimated.

This optimistic estimate has been underscored by a finding of the Hubble Space Telescope in December of 2003. The Space Telescope succeeded in measuring a highly controversial object in an ancient part of our galaxy. It was not known whether this object is a planet or a brown dwarf. It has turned out to be a planet, having two and a half times the mass of Jupiter. It has an estimated age of thirteen billion years, which means that it must have formed in a very young universe, one that was barely one billion years in existence.

Planets keep forming with remarkable speed and abundance to this day. In May of 2004, astronomers trained the new Spitzer Space Telescope at a "star nursery" region of the universe known as RCW 49, and in one image uncovered three hundred newborn stars, some not more than one million years old. A closer look at two of the stars showed that they have faint planet-forming disks of dust and gas around them. The astronomers estimated that all three hundred might harbor such disks. This is a surprising discovery. If planets form around many stars, and if they form so soon, they must be far more abundant than was previously estimated.

If life potentially exists in so many places in the universe, wouldn't intelligent life and even technological civilization also exist? The probabilities in this regard were first calculated by Frank Drake in 1960. The famous Drake equation gives the statistical probabilities of the existence in our galaxy of stars with planets; of planets with environments capable of sustaining life; of life on some of the life-friendly planets; of intelligent life on some of the actually life-bearing planets; and of advanced technological civilization produced by the intelligent life that evolved on these planets. Drake found that, given the large number of stars in our galaxy, as many as ten thousand advanced technological civilizations are likely to exist in the Milky Way galaxy alone.

The Drake equation was updated and elaborated by Carl Sagan and colleagues in 1979. Their computations claim that not ten thousand, but up to one million intelligent civilizations could exist in our galaxy. In the late 1990s, Robert Taormina applied these equations to a region within one hundred light-years from Earth and found that more than eight such civilizations should be present within "hailing distance" from us.

In the last fifteen years, twelve hundred Sun-like stars in our vicinity have been scrutinized by astronomers with ground-based telescopes, and their search has come up with over one hundred extrasolar planets. A particularly promising find was announced in June 2002: the planetary system known as 55 Cancri. It is within hailing distance: forty-one light-years from us. It appears to have a planet that resembles Jupiter in mass and in regard to orbit. Calculations indicate that 55 Cancri could also have rocky planets much like Mars, Venus, and Earth.

However, this is a relatively exceptional find. Most of the other solar systems in our neighborhood have alien planets in widely eccentric orbits, moving either too far from their host sun to sustain life or moving too close to it.

Although planets appear to be highly abundant in this galaxy and elsewhere in the cosmos, planets capable of sustaining advanced forms of life could be relatively rare. According to Peter Ward, radiation and heat levels are so high on most planets that the only forms of life that are likely to exist are a variety of bacteria deep in the soil. The odds against advanced technological civilization beyond Earth, they say, are astronomical. But even if planets with the right composition, the right distance from the host star, and the right orbit were rare in the universe, the existence of advanced civilizations could not be excluded. There are an astronomical number of stars and planets, so even if the odds are astronomically against such civilizations, they do not foreclose their actual existence, but only indicate that they are relatively infrequent.

In light of the finding that planets already started to form a billion years after the birth of the universe, estimates of the prevalence of life in the universe need to be revised upward once again. Even if life-supporting planets are rare and evolution on them is slow, under favorable conditions higher forms of life are likely to have emerged on some planets. Thus extraterrestrial civilizations could well exist in this universe. And some of these civilizations could be more advanced than civilization on Earth: in our region of the galaxy stars that could have life-bearing planets around them are on the average one billion years older than the Sun. Life and civilization could have arisen in this galaxy a billion or more years before they evolved on Earth.

A further factor needs to be added to the estimates regarding the prevalence of life and civilization in the cosmos: the factor of in-formation. In an in-formed universe, the existence of life, and also of advanced civilizations, is far more probable than in a conventional universe. This is because, through the A-field, life in any one place in-forms and facilitates the evolution of life in other places. Evolution never starts from scratch and is not at the mercy of the lucky fluke that random mutations come up with organisms that happen to prove viable in a changing environment.

The evolution of life on Earth did not rely on chance mutations, nor did it require the physical importation of organisms or proto-organisms from elsewhere in the solar system, as the "biological seed-ing" theories of the origins of life suggest. Instead, the chemical soup out of which the first proto-organisms arose was in-formed by the forms of life that had evolved elsewhere in the universe. Life on Earth was not biologically, but *in-formationally* seeded—and its evolution continues to be in-formed by life wherever it exists in the universe.

Can the human brain pick up extraterrestrial information? So-called primitive people have a remarkable faculty of feeling or sensing other people and their environment beyond the range of eye and ear. But we, supposedly civilized people, abandoned this faculty when we relied on our bodily senses to give us information about the world around us. Yet, as shown by our ability to dream, daydream, and receive insights and impressions in meditative and other altered states of consciousness (where the censorship that represses "anomalous" information is lifted), our ability to access a wide range of information has not been lost.

At this crucial juncture in the evolution of human civilization it would be of particular importance to cultivate our long-neglected faculty for accessing the in-formation conserved in the A-field. We would not only develop closer ties to each other and to nature; we might also gain crucial insights into ways to cope with the problems of our technologically evolved but largely rudderless civilization. After all, even if they are statistically rare, a number of technological civilizations are likely to exist in this galaxy, and in the one hundred billion other galaxies of our universe, some of them on planets where life evolved millions if not billions of years before it did on Earth. If these civilizations developed a potent technology, they must also have faced at some point the challenge of finding ways to live with it without damaging their home planet.

The civilizations that met this challenge found ways to achieve a condition of sustainability. What ways did they find? The answer must be in the A-field. Accessing it would be to our advantage: beyond the intrinsic value of knowing that "we are not alone" we could get perhaps vague yet meaningful glimpses of a planetary civilization in harmony with its biosphere. This could make the crucial difference between bumbling along in a fateful gamble with trial and error, and moving with intuitive wisdom toward the dynamically harmonized sustainable conditions that more mature civilizations have already achieved on their home planets.

THE FUTURE OF LIFE IN THE COSMOS

The reasonable certainty that life, even advanced forms of life, exists on other planets does not tell us that life will exist forever. The fact is that life cannot exist indefinitely in the universe: the physical resources required for carbon-based life—the only kind we know of—do not last forever.

The evolution of the known forms of life depends on a strictly limited range of temperatures and the presence of a specific variety of chemical compounds. These factors, as we have seen, are likely to exist on a number of planets in this and other galaxies, on planets that have the right chemical and thermal conditions, situated at the right distance from their active star. But whether such planets are highly abundant or relatively rare, the conditions they provide for the sustenance of life are limited in time. The principal reason is that the active phase of the stars whose radiation drives the processes of life does not last forever. Sooner or later stars exhaust their nuclear fuel, and then they either shrink to the white dwarf stage or fly apart in a supernova explosion. The population of active stars is not infinitely replenished in this universe. Even if new stars keep forming from interstellar dust, a time must come when no further stars are born.

Even if the time dimension is staggering, the limitations are real. About 10^{12} (one trillion) years from now, all the stars that remain in our universe will first have converted their hydrogen into helium—the main fuel of the supercompacted but still luminous white dwarf state—and then will have exhausted their supply of helium. We have already been able to observe that the galaxies constituted of such stars take on a reddish tint, then—when their stars cool still further—fade from sight altogether. As energy is lost in the galaxies through gravitational radiation, individual stars move closer together. The chance of collision among them increases, and the collisions that occur precipitate some stars toward the center of their galaxies and expel others into extragalactic space. As a result, the galaxies diminish in size. Galactic clusters also shrink, and in time both galaxies and galactic clusters implode into black holes. At the time horizon of 10^{34} years, all matter in our universe will be reduced to radiation, positronium (pairs of positrons and electrons), and compacted nuclei in black holes.

Black holes themselves decay and disappear in a process Stephen Hawking called evaporation. A black hole resulting from the collapse of a galaxy evaporates in 10^{99} years, while a giant black hole containing the mass of a galactic supercluster vanishes in 10^{117} years. (If protons do not decay, this span of time expands to 10^{122} years.) Beyond this humanly inconceivable time horizon, the cosmos contains matter particles only in the form of positronium, neutrinos, and gamma-ray photons.

Whether a universe is expanding (open), expanding and then contracting (closed), or balanced in a steady state, the complex structures required for the known forms of life vanish before matter itself supercrunches, or evaporates.

In the late phases of a closed universe—one that ultimately collapses back on itself—the universe's background radiation increases gradually but inexorably, subjecting living organisms to mounting temperatures. The wavelength of radiation contracts from the microwave region into the region of radio waves, and then into the infrared spectrum. When it reaches the visible spectrum, space is lit with an intense light. At that time all stars and planets are vaporized, along with whatever life forms may have already evolved.

In an open universe that expands indefinitely, life dies out because of cold rather than heat. As galaxies continue to move outward, many active stars complete their natural life cycle before gravitational forces bunch them close enough to create a serious risk of collision. But this does not improve the prospects of life. Sooner or later all the active stars of the universe exhaust their nuclear fuel and then their energy output diminishes. The dying stars either expand to the red giant stage, swallowing up their inner planets, or settle into lower luminosity levels on the way to becoming white dwarfs or neutron stars. At these diminished energy levels they are too cold to sustain any form of organic life known to us.

A similar scenario holds in a steady-state universe. As active stars approach the end of their life cycle, their energy output falls below the threshold where life can be supported. Ultimately a lukewarm, evenly distributed radiation fills space, in a universe where the remnants of matter are random occurrences. This universe is incapable of maintaining the flame of a candle, not to mention the complex irreversible reactions that are the basis of life.

Whether our universe expands and then contracts, expands infinitely, or reaches a steady state, the later stages of its evolution will wipe out the known forms of life.

This is a dismal picture, but it is not the whole picture. The whole picture is not limited to our own finite universe; there is also a temporally (whether or not also spatially) infinite or quasi-infinite Metaverse. And life in the Metaverse need not end with the devolution of local universes. While life in each local universe must end, it can evolve again in the universes that follow.

If evolution in each local universe starts with a clean slate, the evolution of life in local universes is a Sisyphean effort: it breaks down and starts again from scratch, time after time. But local universes are not subject to this ordeal. Each universe in-forms the vacuum in which it arose, and its in-formed vacuum in-forms the next universe. Thus in each universe life evolves more and more efficiently, and in equal times evolves further and further, toward coherence and complexity.

Cosmic evolution is a cyclical process with a learning curve. Each universe starts without life, evolves life when some planets become capable of supporting it, and wipes it out when planetary conditions pass beyond the life-supporting stage. But the vacuum shared by all the universes is more and more in-formed, and it creates more and more favorable conditions for the evolution of life.

Cyclically progressive evolution in the Metaverse offers a positive prospect for the future of life: it can continue in one universe after another. And it can evolve further and further, in universe after universe.

What can we say about the super-evolved forms of life that would come about in the mature stages of mature universes? Since the course of evolution is never precisely predictable, we can actually say very little. All we can surmise is that mature organisms in mature universes will be more coherent and complex than the forms of life familiar to us. In most other respects they could be as different from the organisms we know on Earth as humans are different from the protozoan slime that once populated the primeval seas of this planet.

GLIMPSES OF ULTIMATE REALITY

We end the first part of our explorations of the in-formed universe with a question that is meaningful but decidedly not modest: a question about the nature of what mystics and scientists traditionally called "ultimate reality." We have already seen how our universe and possibly myriad other universes in the Metaverse come into being, how they evolve and devolve, and how they give rise to the complex systems we call living. What do these stupendous processes tell us about the ultimate nature of reality?

The answer to this age-old question is now relatively straightforward. *The most fundamental element of reality is the quantum vacuum, the energy- and in-formation-filled plenum that underlies, generates, and interacts with our universe, and with whatever universes may exist in the Metaverse.*

This answer corresponds to an ancient insight: that the universe we observe and inhabit is a product of the energy-sea that was there before there was anything there at all. Hindu and Chinese cosmologies have always maintained that the things and beings that exist in the world are a concretization or distillation of the basic energy of the cosmos, descending from its original source. The physical world is a reflection of energy vibrations from more subtle worlds that, in turn, are reflections of still more subtle energy fields. Creation, and all subsequent existence, is a progression downward and outward from the primordial source.

In Indian philosophy the ultimate end of the physical world is a return to Akasha, its original subtle-energy womb. At the end of time as we know it, the almost infinitely varied things and forms of the manifest world dissolve into formlessness, living beings exist in a state of pure potentiality, and dynamic functions condense into static stillness. In Akasha, all attributes of the manifest world merge into a state that is beyond attributes: the state of *Brahman*.

Although it is undifferentiated, Brahman is dynamic and creative. From its ultimate "being" comes the temporary "becoming" of the manifest world, with its attributes, functions, and relationships. The cycles of *samsara*—of being-to-becoming and again of becoming-to-being—are the *lila* of Brahman: its play of ceaseless creation and dissolution. In Indian philosophy, absolute reality is the reality of Brahman. The manifest world enjoys but a derived, secondary reality and mistaking it for the real is the illusion of *maya*. The absolute reality of Brahman and the derived reality of the manifest world constitute a co-created and constantly co-creating whole: this is the *advaitavada* (the nonduality) of the universe.

The traditional Eastern conception differs from the view held by most people in the West. In the modern conception reality is material. The things that truly exist are bits or particles of matter. They can form into atoms, which can further form into molecules, cells, and organisms—as well as into planets, stars, stellar systems, and galaxies. Matter moves about in space, acted on by energy. Energy also enjoys reality (since it acts on matter), but space does not: space is merely the backdrop or the container in which material things trace their careers.

This typically Western view is a heritage of the Newtonian world-concept. According to Newton, space is a mere receptacle and it is passive in itself; it conditions how things actually behave but does not act on them directly. Although it is empty and passive, Newton maintained that space is nonetheless real: it is an objective element in the universe. Subsequently a number of philosophers, including Gottfried Leibniz and Immanuel Kant, contested the reality of space. In these views, space is nothing in itself; it is merely the way we order relationships among real things. Space itself is not experienced, said Kant; it is only the precondition of experience.

The view that space is empty and passive, and not even real to boot, is diametrically opposed to the view we get at the leading edge of science. What the new physics describes as the unified vacuum—the seat of all the fields and forces of the physical world—is in fact the most fundamentally real element of the universe. Out of it have sprung the particles that make up our universe, and when black holes "evaporate," it is into it that the particles fall back again. What we think of as matter is but the quantized, semistable bundling of the energies that spring from the vacuum.

In the last count matter is but a waveform disturbance in the quasi-infinite energy- and in-formation-sea that is the connecting field, and the enduring memory, of the universe.

SIX

Consciousness—
Human and Cosmic

Next we query the in-formed universe about the nature of consciousness. Did it originate with *Homo sapiens,* or is it part of the fundamental fabric of the cosmos? Will it evolve further in the course of time—and what kind of impact will it have on us and on our children when it does?

We probe deeper still. Could the universe itself possess some form of consciousness, a cosmic or divine root from which our consciousness has grown, and with which it remains subtly connected?

If the in-formed universe is the cornerstone of an integral theory of everything, it should also provide answers to a set of questions centered not on the manifest facts of universe and life, but on the more subtle facts of consciousness.

The questions we pose concern:

- the roots of the phenomenon we know as consciousness
- the wider range of the active in-formation that reaches and forms our (and any other) consciousness
- the next evolution of human consciousness
- the possibility that our consciousness survives the demise of our body

THE ROOTS OF CONSCIOUSNESS

Contrary to a widespread belief, consciousness is not a uniquely human phenomenon. Although we know only human consciousness (indeed, by direct and indubitable experience we know only our *own* consciousness), we have no reason to believe that consciousness would be limited to me and to you, and to other humans.

The kind of evidence that could demonstrate the limitation of consciousness to humans regards the brain: it would be evidence that the human brain has specific features by virtue of which it produces consciousness. Notwithstanding the view advanced by materialist scientists and philosophers that the physical brain is the source of consciousness, there is no evidence of this kind. Clinical and experimental evidence speaks only to the fact that brain function and state of consciousness are correlated, so when brain function ceases, consciousness (usually) ceases as well. We should specify "usually," since there are exceptions to this: as we shall see, in some well-documented cases—among others, those of patients suffering cardiac arrest in hospitals—individuals have had detailed and subsequently clearly recalled experiences during the time their EEG showed a complete absence of brain function.

Functional MRI (magnetic resonance imaging) and other techniques show that when particular thought processes occur, they are associated with metabolic changes in specific areas of the brain. They do *not* show that the cells of the brain that produce proteins and electrical signals also produce sensations, thoughts, emotions, images, and other elements of the conscious mind. How the brain's network of neurons would produce the qualitative sensations that make up our consciousness is beyond the reach of neurophysiological research.

The fact that a high level of consciousness, with articulated images, thoughts, feelings, and rich subconscious elements, is associated with complex neural structures does not tell us that such consciousness is *due* to these structures. The observation that brain function is *associated* with consciousness does not entail that the brain *creates* consciousness.

PHILOSOPHICAL APPROACHES TO THE BRAIN-MIND PROBLEM

The view that consciousness is produced in and by the brain is just one of the many ways philosophically inclined people have envisaged the relationship between the physical brain and the conscious mind. It is the *materialist* way. It maintains that consciousness is a kind of by-product of the survival functions the brain performs for the organism. As organisms become more complex, they require a more complex "computer" to steer them so they can get the food, the mate, and the related resources they need in order to survive and reproduce. At a given point in this development, consciousness appears. Synchronized neural firings and transmissions of energy and chemical substances between synapses produce the qualitative stream of experience that makes up the woof and warp of consciousness. Consciousness is not primary in the world; it is an "epiphenomenon" generated by a complex material system: the human brain.

The materialist way of envisaging the relationship of brain and mind is not the only way. Philosophers have also outlined the *idealist* way. In the idealist perspective, consciousness is the first and only reality; matter is but an illusion created by our mind. This assumption, while outlandish on first sight, makes eminent sense as well: after all, we do not experience the world directly;

we experience it only through our consciousness. We normally assume that there is a qualitatively different physical world beyond our consciousness, but that may be an illusion. Everything we experience could be part of our consciousness. The material world could be merely our invention as we try to make sense of the flow of sensations in our consciousness.

Then there is the *dualist* way of conceiving of the relationship between brain and consciousness, matter and mind. According to dualist thinkers, matter and mind are both fundamental, but they are entirely different, not reducible one to the other. The manifestations of consciousness cannot be explained by the organism that manifests them, not even by the staggeringly complex processes of the human brain. In this view the brain is the seat of consciousness, but it is not identical with it.

In the history of philosophy, materialism, idealism, and dualism were the principal ways of conceiving the relationship between brain and mind. Materialism is still dominant today. Adherence to it poses vexing problems. As the consciousness philosopher David Chalmers put it, the problem it faces is how "something as immaterial as consciousness" can arise from "something as unconscious as matter." In other words, how can matter generate mind? How the brain operates is a comparatively "soft" problem that neurophysiologists will no doubt solve step by step. But the question regarding the way in which "immaterial consciousness" arises out of "unconscious matter" cannot be answered by brain research, for that deals only with "matter," and matter is not conscious. This is the "hard" problem.

Consciousness researchers of the materialist school admit to being greatly perplexed by this problem. The philosopher Jerry Fodor points out that "nobody has the slightest idea how anything material could be conscious. Nobody even knows what it would be like to have the slightest idea about how anything could be conscious." But philosophers who do not take the materialist stance are not disturbed. Peter Russell says that Chalmers' problem is not just hard; it is impossible. Fortunately, Russell adds, it does not need to be solved, for it is not a real problem. We do not need to explain how unconscious matter generates immaterial consciousness, because neither is matter entirely unconscious, nor is consciousness fully divorced from matter.

Russell is right. The neurons in the brain consist of quanta in complex configurations, and quanta are not mere unconscious matter! They stem from the basic constituents of the complex fields that underlie the cosmos and are not devoid of the qualities we associate with consciousness. Leading physicists such as Freeman Dyson and philosophers of the stature of Alfred North Whitehead asserted that even elementary particles are endowed with a form and level of consciousness. "Matter in quantum mechanics," Dyson said, "is not an inert substance but an active agent. . . . It appears that mind, as manifested by the capacity to make choices, is to some extent inherent in every electron." In that case there is no categorical divide between matter and mind.

David Chalmers' "hard" problem evaporates. The rudimentary consciousness of matter at a lower level of organization (the neurons in the brain) becomes integrated in the more evolved consciousness of conscious matter at a higher level of organization (the brain as a whole). This does away with the hard problem of the materialist view without doing the kind of violence to our everyday apprehension of the world that idealism does (according to which all is mind, and nothing but mind). It also does away with the problem of dualism, one that is just a shade less "hard" than that of materialism—because if matter and mind interact (as they must interact in the brain), then we must still say how "something as unconscious as matter" can act on, and be acted on by "something as immaterial as consciousness."

The "hard problem" is no longer a problem, but the question remains: where does the consciousness associated with matter come from? Given that it is not generated by matter (the brain), could it be present in the world independent of matter?

EVOLUTIONARY PANPSYCHISM

The answer to the question above can be formulated in the context of the philosophical position known as *panpsychism*. The adherents of panpsychism claim that psyche—the essence of consciousness—is a universal presence in the world. Both matter and mind—*physis* as well as *psyche*—are omnipresent in the universe. They were present even when the universe was born.

The view from the in-formed universe goes beyond the classical panpsychist view by adding an evolutionary dimension. Psyche is indeed present throughout the universe, but it is not present everywhere in the same way, at the same level of development. Psyche evolves, the same as matter. In the living organisms of this planet they both are relatively highly evolved, and in our species they are the most highly evolved of all. In us human beings psyche is highly articulated: it is our personal consciousness.

Evolutionary panpsychism does not reduce all of reality to structures made up of in-themselves inert and insentient material building blocks (as in materialism), nor does it assimilate all of reality to a qualitative nonmaterial mind (as in idealism). It takes both matter and mind as fundamental elements of reality, but (unlike dualism) does not claim that they are radically separate; they are different aspects of the *same* reality. What we call "matter" is the aspect we apprehend when we look at a person, a plant, or a molecule *from the outside;* "mind" is the aspect we obtain when we look at the same thing *from the inside.*

For each of us the inside view is available only in regard to our own brain. It is not the complex network of neurons that we see when we inspect the felt contents of our brain; what we apprehend is a complex stream of ideas, feelings, intentions, and sensations. But when we inspect anybody else's brain, it is not this stream that we apprehend. What we see is gray matter: a network of neurons firing in complex loops and sequences.

The limitation of the inside view to our own brain does not mean that we alone are conscious and everyone else is but a neurophysiological mechanism operating within a biochemical system. In the panpsychist concept both views—the brain-view as well as the mind-view—are present in all human beings. And not only in all humans, but also in other biological organisms. And not only in organisms, but also in all the systems that arise and evolve in nature, from atoms to molecules, to macromolecules, to ecologies. In the great chain of evolution, there is nowhere we can draw the line, nowhere we could say: below this there is no consciousness, and above there is.

The panpsychist concept has been explored by the philosopher Alfred North Whitehead. In his "organic metaphysics" all things in the world (all "actual entities") have a "physical pole" as well as a "mental pole." Nobel laureate biologist George Wald reached the same conclusion. Mind, he said, rather than emerging as a late outgrowth in the evolution of life, has existed always.

Essentially the same notion was put forward by Apollo astronaut Edgar Mitchell. All things in the world, he said, have a capacity to "know." Less evolved forms of matter, such as molecules, exhibit more rudimentary forms of knowing—they "know" to combine into cells. Cells "know" to reproduce and fight off harmful intruders; plants "know" to turn toward the Sun, birds to fly south in winter. The higher forms of knowing, such as human awareness and intention, have their roots in the cosmos; they were there in potential at the birth of the universe.

We agree. All things in the world—quanta and galaxies, molecules, cells, and organisms—have "materiality" as well as "interiority." Matter and mind are not separate, distinct realities; they are complementary aspects of the reality of the cosmos.

THE WIDER IN-FORMATION OF CONSCIOUSNESS

The in-formed universe gives us a new view of the world, and a new view of life and of mind. It can also give us new answers to another age-old but today frequently asked query about the range of information that can reach our mind: *Do we see the world only through "five slits in the tower"—or can we "open the roof to the sky"?*

The answer is that we can. In the in-formed universe our brain/ mind can access a broad band of information, well beyond the information conveyed by our five sensory organs. We are, or can be, literally "in touch" with almost any part of the world, whether here on Earth or beyond in the cosmos.

When we do not repress the corresponding intuitions, we can be in-formed by things as small as a particle or as large as a galaxy. This, we have seen, is the finding of psychiatrists and psychotherapists who place their patients in an altered state of consciousness and record the impressions that surface in their minds. It was also Mitchell's outer-space experience. In a higher state of consciousness, he remarked, we can enter into deep communication with the universe. In these states the awareness of every cell of the body coherently resonates with what Mitchell identified as "the holographically embedded information in the quantum zero-point energy field."

Access to the A-Field

We can reconstruct how not only sensory, but also nonsensory, information reaches our mind. We have seen that, according to the new physics, the particles and atoms—and the molecules, cells, organisms, and galaxies—that arise and evolve in space and time emerge from the virtual energy sea that goes by the name of quantum vacuum. These things not only originate in the vacuum's energy sea; they continually interact with it. They are dynamic entities that read their traces into the vacuum's A-field, and through that field enter into interaction with each other. A-field traces—the holograms they create—are not evanescent. They persist and in-form all things, most immediately the same kind of things that created them.

This holds true for our body and brain as well. All we experience in our lifetime—all our perceptions, feelings, and thought processes— have cerebral functions associated with them. These functions have wave-form equivalents, since our brain, like other things in space and time, creates information-carrying vortices—it "makes waves." The waves propagate in the vacuum and interfere with the waves created by the bodies and brains of other people, giving rise to complex holograms.

How do the body and the brain "make waves"? Physicists discovered that all things in the universe are constantly oscillating at different frequencies. These oscillations generate wavefields that radiate from the objects that produce them. When the wavefield emanating from one object encounters another object, a part of it is reflected from that object, and a part is absorbed by it. The object becomes energized and creates another wavefield that moves back toward the object that emitted the initial wavefield. The interference of the initial and the response wavefields creates an overall pattern, and this pattern is effectively a hologram. It carries information on the objects that created the wavefields.

Can these holograms be "read out"? We know that in order to derive the information encoded in a hologram, a reference wave is required. It turns out that this wave is always and everywhere available. Peter Marcer has shown that "any waves reverberating through the universe remain coherent with the waves at the source, and are thus sufficient to serve as the reference to decode the holographic information of any quantum hologram emanating from remote locations."

Generations after generations of humans have left their holographic traces in the A-field, and the information in these holograms is available to be read out. The holograms of individuals integrate in a superhologram, which is the encompassing hologram of a tribe, community, or culture. The collective holograms interface and integrate in turn with the super-superhologram of all people. This is the collective in-formation pool of humankind.

We can tune our consciousness to resonate with the holograms in the A-field. The transmission of information in a field of holograms is known: it occurs when the wavefields that make up two (or more) holograms are "conjugate" with each other. The effect is similar to the more familiar effect known as resonance. Tuning forks and strings on musical instruments resonate with other forks and strings that are tuned to the same frequency (or to entire octaves higher or lower than that frequency). The resonance effect is selective: it does not occur when the forks and strings are tuned to a different, unrelated frequency.

The "phase conjugation" that transmits information in holograms is a particular kind of selective resonance. It occurs when two interpenetrating wavefields contain synchronized oscillations at the same frequency. In that event the conjunction of the individual waves creates a spatially and temporally coherent channel of communication between the objects that emit the wavefields. Even when the wavefields contain oscillations at different frequencies, if they are in harmonic resonance (that is, when they constitute a series of two, four, eight, etc., waves per cycle, with synchronized peaks and troughs across the series) they produce a coherent channel of communication. In that case a pathway of nonlocal information-transmission is created across all different scales of organization, from the quantic to the cosmic.

The level and intensity of the transmission of information varies in accordance with the degree of conjugation between the wavefields. It is the most direct and hence evident when the wavefield of one hologram is highly conjugate with the wavefield of another. Lesser conjugation means weaker resonance and lesser effect.

Normally the most direct and evident resonance occurs between our brain and the hologram we ourselves have created. This is the basis of long-term memory. When we remember a thing, a person, or an event from many years ago, or have an intuition that we have already seen or experienced something (so-called *déja-vu* and *déja vecu*), we do not address the memory stores in our brain: we "recall" the information from the hologram that records our experiences.

Such recall could involve more than just our own experiences. Our brain is not limited to resonating with our hologram alone; it can also resonate in the harmonic mode with the holograms of other people, especially with those with whom we have (or had) a physical or emotional bond. The information we get by reading another person's hologram is seldom in the form of explicit words or events; usually it is in the form of intuitions, images, or vague but meaningful sensations. The most widespread and hence familiar among these are the sudden revelatory intuitions of mothers and lovers when their loved ones are hurt or undergo a traumatic experience.

In everyday life our access to the A-field is largely confined to our own hologram. Yet we are not condemned to view the world through five slits in the tower. By entering altered states of consciousness in which our everyday rationality does not filter out what we can apprehend, we can open the roof to the sky. We access a broad range of information that links us to other people, to nature, and to the universe.

THE NEXT EVOLUTION
OF HUMAN CONSCIOUSNESS

Our consciousness is not a permanent fixture: cultural anthropology testifies that it developed gradually in the course of millennia. In the thirty- or fifty-thousand-year history of *Homo sapiens*, the human body did not change significantly, but human consciousness did. It evolved dramatically.

A number of thinkers have attempted to define the specific steps or stages in the evolution of human consciousness. The Indian sage Sri Aurobindo considered the emergence of superconsciousness in some individuals as the next step; in a similar vein the Swiss philosopher Jean Gebser spoke of the coming of four-dimensional integral consciousness, rising from the prior stages of archaic, magical, and mythical consciousness. The American mystic Richard Bucke portrayed cosmic consciousness as the next evolutionary stage of human consciousness, following the simple consciousness of animals and the self-consciousness of contemporary humans.

Ken Wilber's six-level evolutionary process leads from physical consciousness pertaining to nonliving matter energy through biological consciousness associated with animals and mental consciousness characteristic of humans to subtle consciousness, which is archetypal, transindividual, and intuitive. It leads in turn to causal consciousness and, in the final step, to the ultimate consciousness called Consciousness as Such.

Chris Cowan and Don Beck's colorful spiral dynamics, in turn, sees contemporary consciousness evolving from the strategic "orange" stage that is materialistic, consumerist, and success-, image-, status-, and growth-oriented; to the consensual "green" stage of egalitarianism and orientation toward feelings, authenticity, sharing, caring, and community; heading toward the ecological "yellow" stage focused on natural systems, self-organization, multiple realities, and knowledge; and culminating in the holistic "turquoise" stage of collective individualism, cosmic spirituality, and Earth changes.

Ideas such as these differ in specific detail, but they have a common thrust. Consciousness evolution is from the ego-bound to the transpersonal form. If this is so, it is a source of great hope. Transpersonal consciousness is open to more of the information that reaches our brain than the consciousness still dominant today. This could have momentous consequences. It could produce greater empathy among people, and greater sensitivity to animals, plants, and the entire biosphere. It could create subtle contact with the rest of the cosmos. When a critical mass of humans evolves to the transpersonal level of consciousness a higher civilization is likely to emerge, with deeper solidarity and a higher sense of justice and responsibility.

Will such a consciousness-evolution actually come about? This we cannot say: evolution is never fully predictable. But if humankind does not destroy its life-supporting environment and decimate its numbers, the dominant consciousness of a critical mass will evolve from the ego-bound to the transpersonal stage. And this quantum leap in the evolution of consciousness will catalyze a quantum leap in the evolution of civilization as well.

COSMIC CONSCIOUSNESS

We now take another step in our exploration of the in-formed universe: a step that goes beyond the consciousness associated with living organisms. *Could the cosmos itself possess consciousness in some form?*

Through the ages, mystics and seers have affirmed that consciousness is fundamental in the universe. Seyyed Hossein Nasr, a medieval Islamic scholar and philosopher, wrote, "The nature of reality is none other than consciousness. . . . " Sri Aurobindo concurred: "All is consciousness—at various levels of its own manifestation . . . this universe is a gradation of planes of consciousness." Scientists have occasionally joined the ranks of the mystics. Sir Arthur Eddington noted, "The stuff of the universe is mind-stuff . . . the source and condition of physical reality."

Nearly 2,500 years ago Plato advised caution in regard to addressing such ultimate questions: the best we can do is to tell a likely story. We can heed this advice, but can assert in good conscience that the likeliest story is that consciousness extends to the heart of the cosmos:

to the quantum vacuum. We know that this subtle virtual energy sea is the originating ground of the wave-packets we view as matter, and we have good reason to assume that it is the originating ground of mind as well.

How could we tell that the vacuum is not only the seat of a super-dense energy field from which spring the wave-packets we call matter, but also a cosmically extended proto- or root-consciousness? There is no way we could tell by ordinary sensory experience. Consciousness is "private," we cannot ordinarily observe it in anyone other than our-selves. The claim that the vacuum is a field of proto-consciousness, although supported by logical reasoning, is condemned to remain hypothetical.

There are, however, positive approaches we can take. To begin with, even if we cannot directly observe consciousness in the vacuum, we could attempt an experiment. We could enter an altered state of consciousness and identify ourselves with the vacuum, the deepest and most fundamental level of reality. Assuming that we succeed (and transpersonal psychologists tell us that in altered states people can identify with almost any part or aspect of the universe), would we experience a physical field of fluctuating energies? Or would we expe-rience something like a cosmic field of consciousness?

We have noted that when we experience anybody else's brain "from the outside," we do not experience his or her consciousness— at the most we experience a complex set of neurons firing in complex sequences. But when we experience our own brain "from the inside," we experience not neurons, but the qualitative features that make up our stream of consciousness: thoughts, images, volitions, colors, shapes, and sounds. Would not the same hold true when we project ourselves into a "mystical union" with the vacuum?

This is not just a fanciful supposition: there is indirect yet sig-nificant evidence for it. It comes from the farther reaches of contem-porary consciousness research. Stanislav Grof found that in deeply altered states of consciousness, many people experience a kind of consciousness that appears to be that of the universe itself. This most remarkable of altered-state experiences surfaces in individuals who are committed to the quest of apprehending the ultimate grounds of existence. When the seekers come close to attaining their goal, their descriptions of what they regard as the supreme principle of existence are strikingly similar. They describe what they experience as an immense and unfathomable field of consciousness endowed with infinite intelligence and creative power. The field of cosmic consciousness they experience is a cosmic emptiness—a void. Yet, paradoxically, it is also an essential fullness. Although it does not feature anything in a concretely manifest form, it contains all of existence in potential. The vacuum they experience is a plenum: nothing is missing in it. It is the ultimate source of existence, the cradle of all being. It is pregnant with the possibility of everything there is. The phenomenal world is its creation: the realization and concretization of its inherent potential.

Basically the same kind of experience is recounted by people who practice yoga and other forms of deep meditation. The Indian Vedic tradition, for example, regards consciousness not as an emergent property that comes into existence through material structures such as the brain and the nervous system, but as a vast field that constitutes the primary reality of the universe. In itself, this field is unbounded and undivided by objects and individual experiences, but it can be experienced by individuals in meditation when the gross layers of the mind are stripped away. Underlying the diversified and localized gross layers of ordinary consciousness there is a unified, nonlocalized, and subtle layer: "pure consciousness."

Thus the evidence for cosmic consciousness is not entirely indirect: it has an experiential basis. Combining the implications of the in-formed universe with the testimony of altered-state experiences we can state—more exactly, *re*-state—the likeliest story.

The original statement is thousands of years old. According to the ancient cosmologies the universe's undifferentiated, all-encompassing consciousness separates off from its primordial unity and becomes localized in particular structures of matter. This insight can be restated in the context of cutting-edge science. In that context we specify that the proto-consciousness that infuses the cosmos becomes localized and articulated as particles emerge from the vacuum and evolve into atoms and molecules. On life-bearing planets the atoms and molecules evolve into cells, organisms, and ecologies. Through them the consciousness that infuses the cosmos becomes more and more articulated. The human mind, associated with the remarkably evolved human brain, is the highest-level articulation on this planet of the consciousness that, arising from the vacuum, pervades the cosmos.

THE FARTHERMOST REACHES OF CONSCIOUSNESS

Last but not least we ask perhaps the most exciting of all the great questions people have ever asked. *Could our consciousness survive the physical demise of our body?*

Consciousness Beyond the Brain

If we are to shed light on the perennial question of the survival of consciousness, we must venture beyond the observational methods of the natural sciences. It does not help to observe the human brain, for if consciousness continues to exist when brain function ceases, it is no longer associated with the brain. It is more to the point to look at the evidence furnished by instances where consciousness is no longer directly linked with the brain. This is the case in near-death experiences (NDEs), out-of-body experiences (OBEs), past-life experiences, some varieties of mystical and religious experiences, and, perhaps most significant of all, the experiences of after-death communication (ADC). Until recently, scientists could not cope with such "paranormal" experiences; they did not fit into the materialist scheme of scientific thinking. But this is not a materialist kind of universe, and in it consciousness is not produced by, and is not limited to, matter.

We have clinical evidence that consciousness can persist in the total absence of brain activity. Cardiologist Pim van Lommel studied near-death experiences of survivors of cardiac arrest in ten Dutch hospitals. He conducted a standardized interview with sufficiently recovered patients within a few days of resuscitation, and asked whether they could remember the period of unconsciousness, and what they recalled. He coded the experiences reported by the patients according to a weighted index. Van Lommel found that 282 of 344 patients had no recollection of the period of cardiac arrest. But 62 reported some recollection of what happened during the time they were clinically dead; of these, 41 had a deep NDE.

A study carried out by B. Greyson in the United States involved 116 survivors of cardiac arrest. Eighteen of the patients reported memories from the period of cardiac arrest; of these, seven reported a superficial experience and eleven had a deep NDE. Greyson wrote, "The paradoxical occurrence of heightened, lucid awareness and logical thought processes during a period of impaired cerebral perfusion raises particular perplexing questions for our current understanding of consciousness and its relation to brain function. A clear sensorium and complex perceptual processes during a period of apparent clinical death challenge the concept that consciousness is localized exclusively in the brain."

British researchers Sam Parnia and Peter Fenwick concurred. The data suggest, they wrote, that near-death experiences do arise during unconsciousness. This is a surprising conclusion, they added, because when the brain is so dysfunctional that the patient is deeply comatose, the cerebral structures, which underpin subjective experience and memory, must be severely impaired. Complex experiences should not arise or be retained in memory.

Van Lommel concluded, "Our waking consciousness is only a part of our whole undivided consciousness. There is also an extended or enhanced consciousness based on indestructible and constant evolving fields of information, where all knowledge, wisdom and unconditional love are present and available."

Reincarnation

Consciousness, it appears, can persist in the absence of brain function. Does this mean that it can reappear in the body and brain of another person? Let us take an unbiased look at the pertinent phenomena.

Phenomena suggestive of reincarnation consist of impressions and ideas recounted by people about sites, persons, and events they have not and could not have encountered in their present lifetime. Such phenomena crop up routinely in the experience of psychotherapists who practice regression analysis. In this therapeutic process the therapists place their patients in a slightly altered state—hypnosis is not needed, since breathing exercises, rapid eye movements, or simple suggestion is usually sufficient—and take them back from their current experiences to the experiences of their past. They can often move their patients back to early childhood, infancy, and physical birth. Experiences that seem to be those of gestation in the womb surface as well.

Interestingly, and at first quite unexpectedly, the therapists have found that they can take their patients back beyond the womb and physical birth. After an interval of apparent darkness and stillness, other experiences appear. They are of other places and other times. Yet the patients not only recount them as the experience of a book they have read or a film they have seen, but actually *re-live* them. As Stanislav Grof's records testify, they become the person they experience, even to the inflection of voice, the language (which may be one the patient has never known in his or her present lifetime), and, if the experience is of infancy, the involuntary muscle reflexes that characterize infants.

Ian Stevenson of the University of Virginia investigated the past-life experiences recounted by children. During more than three decades Stevenson interviewed thousands of children, in both the West and the East. He found that from the age of two or three, when they begin to verbalize their impressions, until the age of five or six, many children report identification with people they have not seen, heard of, or encountered in their young lives. Some of these reports have been verified as the experience of a person who had lived previously, and whose death matched the impressions reported by the child. Sometimes the child carried a birthmark associated with the death of the person with whom she or he identified, such as an indentation or discoloration on the part of the body where a fatal bullet entered, or malformations on the hand or foot the deceased had lost or had wounded. One child in India, called Parmod, recalled in detail a previous life in a neighboring village and could identify people and places there with great precision and remarkable detail.

Parmod's story is not unique. There is a wide array of evidence for past-life experiences, but the evidence does not guarantee a correct interpretation. Spiritually disposed persons tend to assume that these experiences are from a previous life, yet this is merely one interpretation. The interpretation that is more consistent with what we know of the in-formed universe is that our brain becomes tuned to the holographic record of another person in the vacuum. "Past-life experiences" signify the retrieval of information from the A-field, rather than the incarnation of the spirit or soul of a dead person.

Immortality

If all that we experience enters the A-field we have a cogent explanation of experiences that appear to stem from past lives. But there is another variety of anomalous experience that calls for explanation as well: it is the experience of *communicating* with a recently deceased person. We are dealing here not with re-living someone else's experience as our own, but with encountering someone else after he or she has died. The deceased does not appear to us as a set of experiences that could stem from our own past existence, but as another person who is still alive in some sense, for he or she can communicate with us.

Once again, let us review the pertinent evidence. In near-death experiences, out-of-body experiences, past-life experiences, and various mystical and religious experiences, people seem to perceive things that were not conveyed by their eyes, ears, or other bodily senses. In

NDEs the brain can be clinically dead, with the EEG "flat," and yet people can have clear and vivid experiences that, when they come back from the portals of death, they can recall in detail. In OBEs people can "see" things from a point in space that is removed from their brain and body, while in mystical and religious transport experiencing subjects have the sense of entering into union with something or someone larger than themselves, and perhaps larger or higher than the natural world. Although in some of these experiences the consciousness of individuals is detached from their physical brain, their experiences are vivid and realistic. Those who undergo them seldom doubt that they are real. But communicating with persons who are deceased is another variety of experience; it implies not the incarnation of one person's spirit or soul in the body of another, but the persistence of his or her spirit or soul independent of the body. If true, this would indicate some form of *immortality*.

Many people seem to experience ADCs—instances of after-death communication. Mediums such as James Van Praagh, John Edward, and George Anderson have mediated contact with thousands of deceased people by describing the impressions they receive from them. NDE researcher Raymond Moody collected a wide variety of "visionary encounters with departed loved ones."

ADCs often occur spontaneously, but they can also be induced. Allan Botkin, a qualified psychotherapist, head of the Center for Grief and Traumatic Loss in Libertyville, Illinois, and colleagues claim to have successfully induced after-death communication in nearly three thousand patients. ADCs can be induced in about 98 percent of the people who try them. Usually the experience comes about rapidly, almost always in a single session. It is not limited or altered by the grief of the subject or his or her relationship to the deceased. It also does not matter what the experiencers believed prior to undergoing the experience, whether they were deeply religious, agnostic, or convinced atheists.

ADCs can occur also in the absence of a personal relationship with the deceased—for example, in combat veterans who feel grief for an anonymous enemy soldier they have killed. And they can occur without guidance by the psychotherapist. Indeed, as Dr. Botkin reports, leading the experiencing subject actually inhibits the unfolding of the experience; it is sufficient to induce the mental state necessary for the experience to occur. This state is a slightly altered state of consciousness, brought about by means of a series of rapid eye movements. Known as "sensory desensitization and reprocessing," it produces a receptive state in which people are open to the impressions that appear in their consciousness.

Typically, the experience of after-death communication is clear, vivid, and thoroughly convincing. The therapists hear their patients describe communication with the deceased person, hear them insist that their reconnection is real, and watch repeatedly as their patients move almost instantly from an emotional state of grieving to a state of relief and elation.

The story of a young man who inadvertently killed a couple and their daughter by straying onto the wrong lane on the highway is particularly illuminating. Mark was not injured, but his life changed from that day; he awoke each morning to deep sadness and severe guilt. He twice attempted suicide, had two failed marriages, and was on the verge of losing his job. Then he underwent an induced ADC experience. Botkin reported that Mark sat quietly, with closed eyes. After a moment he said, "I can see them. It's the family with the little girl. They're standing together and smiling. . . . Oh God, they look happy and peaceful. They're very happy being together and they're telling me they very much like where they are." Mark continued: "I can see each one very clearly, and especially the girl. She's standing in front of her mom and dad. She has red hair, freckles, and a wonderful smile. I can see the dad walking around, like he's showing me how he can walk. I have the feeling from him that he had multiple sclerosis before he died, and he is really happy he can now move around freely." (As subsequent inquiry showed, the father did indeed have multiple sclerosis at the time he died.) Mark told the family that he was very sorry about what had happened and heard them say that they forgave him. He felt as if a huge burden had been lifted from him.

This experience is fairly typical. In ADCs people experience the person they grieve for as happy and well, often younger than they were at the time they died. A "reconnection" with the deceased relieves and often resolves the grief weighing on the mind of the experiencer.

Clearly, ADCs have remarkable therapeutic value. But what do they mean? Are they grief-induced delusions? Botkin argues that they are not; they do not fit any of the known categories of hallucinations. Then are they real: do the subjects actually encounter the deceased for whom they are grieving? That would suggest that the deceased still exists in some way, perhaps in another dimension of reality. This would be true immortality: the survival of the person—the consciousness, spirit, or soul of the person—after the physical demise of his or her body.

The American philosopher Chris Bache, who for more than twenty years experimented with deeply altered states of consciousness, wrote to the author:

> In my inner work, wherever I touched my life in nonordinary states, it broke open to reveal a tapestry of collective threads. I could not find any part of "my" existence that was not part of the larger tapestry of life. And yet, over the course of years, things happened that seemed to suggest that something was being birthed in these experiences that would endure beyond any frame of reference previously imaginable to me, beyond egoic existence, beyond any space-time structure altogether. I found it necessary to affirm the emergence of a new and higher form of individuality being generated by the universe's relentless accumulation of experience—both accumulation through many cycles of reincarnation, and through the systematic integration into one point of awareness of vast territories of transpersonal experience.

Gustav Fechner, the pragmatic founder of experimental methods in psychology, came to much the same conclusion. "When one of us dies," he wrote after recovering from a serious illness, "it is as if an eye of the world were closed, for all perceptive contributions from that particular quarter cease. But the memories and conceptual relations that have spun themselves round the perceptions of that person remain in the larger Earth-life as distinct as ever, and form new relations and grow and develop throughout all the future, in the same way in which our own distinct objects of thought, once stored in memory, form new relations and develop throughout our whole finite life."

The mystic Alice Bailey identified the "larger Earth-life" in terms consistent with the in-formed universe. "This word 'ether'," she wrote, "is a generic term covering the ocean of energies which are all interrelated and which constitute that one synthetic energy body of our planet . . . the etheric or energy body, therefore, of every human being is an integral part of the etheric body of the planet itself."

What can we conclude from all these nonordinary experiences and the evidence they provide?

A Last Reflection

There is much that we do not yet understand about the farthermost reaches of human consciousness, but one thing stands out: consciousness does not vanish when the functions of the brain and body cease. It persists, can be recalled and, for a time at least, can also be communicated with. It appears that the hologram that codes the experiences of a lifetime maintains a level of integration that allows it a form of autonomous existence even when it is no longer associated with a brain and a body. It is capable of receiving inputs from the manifest world and of responding to them. In this interpretation, the perennial intuition of an immortal soul is no longer inconsistent with what we are now beginning to comprehend through science about the true nature of reality.

SEVEN

The Poetry of Akashic Vision

The universe is a memory-filled world of constant and enduring interconnection, a world where everything in-forms—acts on and interacts with—everything else. We should apprehend this remarkable world with our heart as well as with our intellect. This chapter speaks to our heart. It recalls the ancient intuition of an information-filled cosmos where everything is conserved and everything affects everything else. It offers a vision that is *imaginative* but not *imaginary*: a poetic vision of a cosmos where nothing disappears without a trace, and where all things that exist are, and remain, intrinsically and intimately interconnected.

The vision of a coherent, interconnected, and cyclically self-renewing cosmos is not new. The most important among its historical antecedents is the vision that inspired the imagination of countless generations in India and throughout the East: the vision of a world that emerged from Akasha.

Akasha can be described rationally, in terms of cutting-edge science, but it can also be depicted poetically. A poetic description is important, because if a world where the Akashic Field connects everything with everything else is the best insight we have into the fundamental nature of reality, we should not only grasp it with our intellect: we should also allow it to resonate with our hearts and inform our dreams.

Here, then, is the Akashic vision of the birth and rebirth of our universe, addressed not to our intellect, but to our heart.

A lightless, soundless, formless plenum. It is filled both with the primeval consciousness that is the womb of all mind and spirit in the cosmos and with the fluctuating energies out of which all things come to exist in space and in time. There is no-thing in this cosmic fullness, yet there is every-thing, in potential. Everything that can and will ever happen is here, in formless, soundless, lightless, quiescent turbulence.

After an infinity of cosmic eons, a sudden explosion, untold magnitudes greater than any turbulence ever witnessed or even imagined by human beings, penetrates the formless turbulence; a shaft of light rises from its epicenter. The plenum is no longer quiescent; it is rent by a supercosmic force emerging from its hitherto soundless and lightless depth. It liberates gigantic forces, transforming the plenum from virtual formlessness into dynamic formative process. The surface foams with instantly appearing and disappearing ripples of energy, forming and annihilating in a cosmic dance of unimaginable speed and momentum. Then the initial demented rhythm becomes more sedate, the foam more orderly. The ripples radiate outward from the epicenter, bathed in pure light of infinite intensity.

As the foam expands, it becomes grainier. Swirls and vortexes appear, incipient if as yet evanescent wave-patterns modulating the surface of the evolving plenum. With the passing of further cosmic eons, the ripples of patterned energy consolidate into lasting forms and structures. They are not separate from each other, for they are micropatterns structuring into larger patterns within a common wavefield. They are part of the underlying and now no longer formless plenum that erupted and created them. Each ripple is a microworld in itself, pulsating with the liberated energies of the plenum and reflecting in its micrototality the macrototality from which it emerged.

The micropatterns trace their careers in the expanding space of the initial explosion and take on structure and complexity. They modulate the turbulent plenum. It is more and more structured at the surface, as the ripples cohere into complex wave-structures; and it is more and more modulated below, as the evolving structures create minute vortices that integrate into information-carrying holograms. The in-formed holofield below and the micropatterns on the surface evolve together. Their growing architecture enriches the holofield, and the enriched holofield in-forms the evolving microstructures. Surface and depth coevolve, taking on complexity and coherence.

The more complex the structures that emerge, the more independent they appear of the depth below. Yet the ripples and waves at the surface are not separate but part of the medium from which they arise—they are like "solitons," the curiously objectlike waves that emerge in a turbulent medium.

The ripples and waves cohere in elaborate structures, subtly interconnected with each other. At a crucial stage of their evolution they become self-sustaining, reproducing themselves and replenishing spent energies from the embedding energy fields.

The evolving wave-patterns have not just external relations; they also have an inner reflection: a "feel" of each other and of the depth. At first an unarticulated basic sensation, this inner reflection gains in articulation as the self-maintaining waves acquire structure and complexity. They develop higher and higher grades of inner reflection, articulating their basic feel of the world as a representation of individual things and processes. They map the world that envelops them, and themselves in that world.

After another cosmic eon, the energies liberated by the initial explosion dissipate across the surface of the plenum. Some megastructures use up the free energies available to them and explode, strewing their microripples into space where they consolidate into new megastructures. Others implode, and in a final flash reenter the plenum from which they emerged. The ripples that evolve on the surface of smaller megastructures break down, incapable of maintaining themselves in an environment of fading energy. As the universe ages, all complex structures and articulated reflections disappear. But although the surface loses modulation, the memory of the depth is not affected: the holograms created by the ripples remain untouched. They conserve the trace of the surface's evanescent structures together with their feels and reflections.

And now another shaft of light rends the plenum, breaking its quiescent turbulence and reviving it with another formative burst: a new universe is born. This time the ripples and structures that form on the surface do not appear randomly, at the mercy of chance: they derive from a plenum in-formed with the holo-trace of prior ripples and waves.

The cosmic drama repeats time after time. Further shafts of light radiate outward from the epicenter, another multitude of ripples moves outward to dance, to cohere, to feel, and to reflect. The new universe ends as the ripples and the structures it brought into being vanish at the surface. But the holograms created by them in the depth inform the next universe, born as further explosions rend the plenum. Time after time, the cosmic drama repeats, but it does not repeat in the same way. It builds on its own past, on the memory of the ripples and waves that appeared and then disappeared in prior universes.

In universe after universe the plenum brings forth microripples and megawave structures. In each universe the ripples and waves vanish, but their memory lives on. In the next universe new and more elaborate structures appear, with more articulated reflections of the world around them.

In the course of innumerable universes, the pulsating Metaverse realizes all that the primeval plenum held in potential. The plenum is no longer formless: its surface is of unimaginable complexity and coherence; its depth is fully in-formed. The cosmic proto-consciousness that endowed the primeval plenum with its universe-creative potentials becomes a fully articulated cosmic consciousness—it becomes, and thenceforth eternally is, THE SELF-REALIZED MIND OF GOD.

The Phenomenon of Coherence

A Deeper Look at the Scientific Evidence

The integral theory of everything developed in this book rests on the premise that the nonlocal forms of coherence discovered in the various domains of investigation can be traced to a specific kind of information, namely, "in-formation." The field that records and conveys in-formation in nature, we have said, is the A-field. We now take a deeper look at the scientific evidence for nonlocal coherence—which is at the same time evidence for a field that is responsible for it. We review the phenomenon of coherence in the world of the quantum, in the universe at large, in the domains of life, as well as in the sphere of human consciousness.

COHERENCE IN THE QUANTUM WORLD

Quantum nonlocality. Newton's mass points and Democritus's atoms could be unambiguously defined in terms of force, position, and motion, but the quantum cannot. As we have seen, its description is complex and intrinsically ambiguous. The quanta of light and energy that emerge in sophisticated experiments do not behave as tiny equivalents of familiar objects. In fact, their behavior proves more and more weird. Though Einstein received the Nobel Prize for his work on the photoelectric effect (where streams of light quanta are generated on irradiated plates), he did not suspect—and was never ready to accept—the strangeness of the quantum world. But physicists investigating the behavior of these packets of light and energy have found that, until an instrument of detection or another act of observation registers them, they do not have a specific position, nor do they occupy a unique state. The ultimate units of physical reality have no uniquely determinable location, and they exist in a strange state that consists of the simultaneous "superposition" of several ordinary states.

Until very recently (for evidence contrary to this assumption has now surfaced), quanta were believed to exhibit the property Niels Bohr called "complementarity." Depending on how they were observed and measured, particles were said to be either corpuscles or waves, but not both at the same time. The alternative properties of particles were held to be complementary: although they do not appear singly, together they fully describe the state of the particles. Moreover, as Heisenberg's "principle of uncertainty" specifies, the various states of quanta cannot all be measured at the same time. If one measures position, for example, then momentum (which is the product of mass and velocity) becomes indistinct; and if one measures momentum, position becomes blurred.

Even more weird is the finding that—until it is measured or interacted with in some way—a quantum exists in a state in which all of its possible real states are superposed. Schrödinger's wave function relates the quantum's superposed wave state to its real state. (A "real" state is a classical state, with unique location and normal measurability.) However, there are no laws of physics that can predict which of its possible real states the particle will select. While in the aggregate the shift from the virtual into the real state conforms to statistical rules of probability, there is no way to tell just how it will occur in a given instance. Unless each shift takes place in a separate universe (as Everett suggested in his "parallel universes" hypothesis), individual quantum jumps are indeterminate, not subject to any law of physics.

Einstein was opposed to the fundamental role of chance in nature—he said: "God doesn't play dice." Something is missing in the observational and theoretical arsenal of quantum mechanics, he suggested; in some essential respects the theory is incomplete. Bohr countered that the very question of what a particle is "in-itself" is not meaningful and should not even be asked. The Nobel laureate physicist Eugene Wigner echoed this view: he said that quantum physics deals with *observations,* and not with *observables.* Heisenberg also supported it when he spoke of the error of the "philosophical doctrine of Democritus," which claims that the whole world is made up of objectively existing material building blocks called atoms. The world, said Heisenberg, is built as a mathematical, and not as a material, structure. In consequence there is no use asking to what the equations of mathematical physics refer—they do not refer to anything beyond themselves.

Other physicists, among them David Bohm, refused to accept the quantum physical concept as a full description of reality. His "hidden variables theory" suggests that the selection of the state of the quantum is not random; it is guided by an underlying physical process. In Bohm's theory a pilot wave, identified as the quantum potential "Q," emerges from a deeper, unobservable domain of the universe and guides the observed behavior of particles. Thus, particle behavior is indeterministic only at the surface; at the deeper level it is determined by the quantum potential. Later Bohm called the deeper level of reality the "implicate order," a holofield where all the states of the quantum are permanently coded. Observed reality is the "explicate order;" it is rooted in, and unfolds from, the implicate order.

Various versions of Bohm's theory are being developed today by theoretical physicists who are unwilling to take the mathematical formalisms of quantum physics for an adequate explanation of the real world. They account for the state of the quantum in reference to its interaction with the quantum vacuum, the deep dimension of the universe that has replaced the "luminiferous ether" of the nineteenth century.

This is a relatively recent development. Until the 1980s, quantum weirdness was generally accepted as an irreducible condition of the ultrasmall domain of the universe. Physicists contented themselves with the smooth functioning of the equations by which they computed their observations and made predictions. In the last two decades the picture has begun to change. With the new experiments a far less weird view of the quantum world is beginning to take shape. Experiments that were originally designed to investigate the complementary corpuscular/wave nature of the quantum have been instrumental in bringing it about.

The first of the relevant experiments was conducted by Thomas Young in 1801. In his famous "double-slit experiment," coherent light is allowed to pass through a filtering screen with two slits. (Young created coherent light by making a ray of sunlight penetrate a pinhole; today, lasers are used for this purpose.) When Young placed a second screen behind the filter with the two slits, he found that instead of two pinpoints of light, a wave interference pattern appears on the screen. The same effect can be observed on the bottom of a pool when two drops or pebbles disturb the sunny and otherwise smooth surface of the water. The waves spreading from each disturbance meet and interfere with each other: where the crest of one wave meets the crest of the other, they reinforce each other and appear bright. Where crest meets trough they cancel each other and appear dark.

Are the quanta that pass through Young's slits waves? If so, they could then pass through both slits and form interference patterns. This assumption makes sense until such a weak light source is used in the experiments that only one photon is emitted at a time. Commonsense reasoning tells us that a single photon cannot be a wave: it must be a corpuscular packet of energy of some sort. But then it should be able to pass through only one of the slits and not both slits at the same time. Yet when single photons are emitted, a wave interference pattern builds up on the screen, as if each photon passed through both slits.

The "split-beam experiment," designed by John Wheeler, discloses the same dual effect. Here, too, photons are emitted one at a time, and they are made to travel from the emitting gun to a detector that clicks when a photon strikes it. A half-silvered mirror is inserted along the photon's path, which splits the beam. This means that on the average, one in every two photons will pass through the mirror and one in every two will be deflected by it. To verify this, photon counters are installed both behind the half-silvered mirror and at right angles to it. There is no problem here: the two counters register an approximately equal number of photons. But a curious thing occurs when a second half-silvered mirror is inserted in the path of the photons that are undeflected by the first. One would still expect that an equal number of photons would reach the two counters: deflection by the two mirrors would simply have exchanged their individual destinations. But this is not the case. One of the two counters registers all the photons—none arrives at the other.

It appears that the kind of interference that was noted in the double-slit experiment occurs in the split-beam experiment as well, indicating that individual photons are behaving as waves. Above one of the mirrors the interference is destructive (the phase difference between the photons is 180 degrees), so that the wave patterns of the photons cancel each other. Below the other mirror the interference is constructive (since the wave phase of the photons is the same) and in consequence the photon waves reinforce each other.

The interference patterns of photons emitted moments apart in the laboratory has also been observed in photons emitted at considerable distances from the observer, at considerable intervals of time. The "cosmological" version of the split-beam experiment bears witness to this. In this experiment the photons are emitted not by an artificial light source, but by a distant star. In one case the photons of the light beam emitted by the double quasar known as 0957+516A,B were tested. This distant "quasi-stellar object" appears to be two, but is in fact one and the same object, its double image being due to the deflection of its light by an intervening galaxy situated about one fourth of the distance from Earth. (The presence of mass, according to relativity theory, bends space and hence also curves the path of the light beams that propagate in it.) A light beam taking the curved path takes longer to travel than one coming by the straight path. In this case the additional distance traveled by the light deflected by the intervening galaxy means that the photons that make up the deflected beam have been on the way fifty thousand years longer than those that traveled by the more direct route. Although originating billions of years ago and arriving with an interval of fifty thousand years, the photons of the two light beams interfere with each other just as if they had been emitted seconds apart in the same laboratory.

Repeatable and indeed oft repeated experiments show that particles that originate from the same source interfere with each other, whether they are emitted at intervals of a few seconds in the laboratory or at intervals of thousands of years in the universe. How is this possible? Is a photon or an electron a corpuscle when emitted (since it can also be emitted one by one) and a wave when it propagates (since it produces wavelike interference patterns when it encounters other photons or electrons)? And why does the coupling of this particle/wave persist even over cosmological distances? The search for an answer to these questions points in a new direction.

Recent versions of the double-slit experiment furnish an indication of the direction in which the answer is now being sought. Initially the experiments were designed to answer a simple question: Does the particle really pass through both slits or only through one? And if only one, which one? The experiment consists of an apparatus that allows each photon access to only one of the two slits. When a stream of photons is emitted and confronted with the two slits, the experiment should decide which of the slits a given photon is passing through.

In accordance with Bohr's principle of complementarity, when the experiment is set up so that the path of the photons can be observed, the corpuscular face of the photons appears and the wave-face disappears: the interference fringes diminish and can entirely vanish. (This, we should note, does not mean that the wave-aspect is not present, only that it is not registered by this particular experimental apparatus.) The higher the power of the "which-path detector," the more the interference fringes diminish. This was shown by an experiment conducted by Mordehai Heiblum, Eyal Buks, and colleagues at Israel's Weizmann Institute. Their state-of-the-art technology comprised a device less than one micrometer in size, which creates a stream of electrons across a barrier on one of two paths. The paths focus the electron streams and enable the investigators to measure the level of interference between the streams. The higher the detector is tuned for sensitivity, the less there is of interference. With the detector turned on for both paths, the interference fringes disappear.

This result appears to bear out Bohr's theory, according to which the two complementary faces of particles can never be observed at one and the same time. However, an ingenious experiment by Shahriar Afshar, a young Iranian-American physicist, demonstrated that even when the corpuscular face is observed, the wave-aspect is still there: the interference pattern does not disappear. In this experiment, reported in July 2004 by the British journal *New Scientist,* a series of wires are placed precisely where the dark fringes of the interference pattern should be. When light hits the wires, they scatter it so that less light reaches the photon detector. But light does not reach these particular points: even when photons pass through the slits one at a time, the dark fringes are still in place.

The continued presence of the interference pattern suggests that particles continue to behave as waves even when they are individually emitted, but in that case their wave-face is not apparent to conventional observation. Afshar suggests—and a number of particle physicists are inclined to agree—that the wave aspect of the particle is the fundamental aspect. The corpuscular face is not the real face: the entire experiment can be described in terms of photon-*waves.*

Does this mean that the mysteries surrounding the behavior of particles are resolved? Not by any means. Even as a wave state, the state of the particle is decidedly non-commonsensical: it is "nonlocal." The "which-path detecting apparatus" appears to be coupled in a nonlocal manner with the photons passing through the slits. The effect is astonishing. In some experiments the interference fringes disappear as soon as the detector apparatus is readied—and even when the apparatus is not turned on! Leonard Mandel's optical interference experiment of 1991 bears this out. In the Mandel experiment two beams of laser light are generated and allowed to interfere. When a detector is present that enables the path of the light to be determined, the interference fringes disappear as Bohr predicted. But the fringes disappear regardless of whether or not the determination is actually carried out. The very possibility of "which-path-detection" destroys the interference pattern.

This finding was confirmed in the fall of 1998, when University of Konstanz physicists Dürr, Nunn, and Rempe reported on an experiment where interference fringes are produced by the diffraction of a beam of cold atoms by standing waves of light. When no attempt is made to detect which path the atoms are taking, the interferometer displays fringes of high contrast. However, when information is encoded within the atoms as to the path they take, the fringes vanish. The labeling of the paths does not need to be read out to produce the disappearance of the interference pattern; it is enough that the atoms are labeled so that this information can be read out.

Is there an explanation for this strange finding? There is. It appears that whenever one encodes "directional information" in a beam of atoms, this information correlates the atom's momentum with its internal electronic state. Consequently when an electronic label is attached to each of the paths the atoms can take, the wave function of one path becomes orthogonal—at right angles—to the other. And streams of atoms or photons that are orthogonal cannot interfere with each other.

The fact is that atoms, the same as particles, are nonlocally correlated with each other, and can also be nonlocally correlated with the apparatus by which they are measured.

In itself, the finding of instant connections in the quantum world is not new: "quantum nonlocality" has been known for more than half a century. Already in 1935 Erwin Schrödinger suggested that particles do not have individually defined quantum states but occupy collective states. The collective superposition of quantum states applies not only to two or more properties of a single particle, but also to a set of particles. In each case it is not the property of a single particle that carries information, but the state of the ensemble in which the particle is embedded. As the particles are intrinsically "entangled" with each other, the superposed wave function of the entire quantum system describes the state of each particle within it.

The mutual entanglement of quanta indicates that information is subtly but effectively transmitted throughout the quantum world. As this informational linking is both instant and enduring, it appears to be independent of space as well as of time.

COHERENCE IN THE UNIVERSE

The coherence of cosmic ratios. As we have seen, the observed parameters of the universe are surprisingly coherent. In the 1930s, Sir Arthur Eddington and Paul Dirac noted that "dimensionless ratios" relate the universe's basic parameters to each other. For example, the ratio of the electric force to the gravitational force is approximately 10^{40}, and the ratio of the observable size of the universe to the size of elementary particles is likewise around 10^{40}. This is all the more strange as the former ratio should be unchanging (the two forces are assumed to be constant), whereas the latter is changing (since the universe is expanding). In his "large number hypothesis," Dirac speculated that the agreement of these ratios, the one variable, the other not, is not merely a temporary coincidence. But if the coincidence is more than temporary, either the universe is not expanding or the force of gravitation varies in accordance with its expansion!

Additional coincidences involve the ratio of elementary particles to the Planck-length (this ratio is 10^{20}) and the number of nucleons in the universe ("Eddington's number," which is approximately 2×10^{79}). These are very large numbers, yet "harmonic" numbers can be constructed from them. For example, Eddington's number is roughly equal to the square of 10^{40}.

Menas Kafatos and Robert Nadeau showed that many of these coincidences can be interpreted in terms of the relationship on the one hand between the masses of elementary particles and the total number of nucleons in the universe, and on the other between the gravitational constant, the charge of the electron, Planck's constant, and the speed of light. Scale-invariant relationships appear—the physical parameters of the universe turn out to be extraordinarily correlated and coherent.

The "horizon problem." The coherence implied by numerical relationships is reinforced by observational evidence. The latter gives rise to the so-called horizon problem: the problem of the large-scale uniformity of the cosmos at all points of the horizon as seen from Earth.

The universe's microwave background radiation proves to be isotropic—the same in all directions. This radiation is believed to be the remnant of the Big Bang; according to BB theory, it was emitted when the universe was about 400,000 years old. The problem is that at that point in time the opposite sides of the expanding universe were already ten million light-years apart. By that time, light could have traveled only 400,000 light-years, so no physical force or signal could have connected regions ten million light-years distant. Yet the cosmic background radiation is uniform for billions of light-years wherever we look in space.

This is true not only of the background radiation; galaxies and multigalactic structures also evolve in a uniform manner in all directions from Earth. This is the case even in regard to galaxies that have not been in physical contact with each other since the birth of the universe. Whether it is 13.7 billion or, as the latest findings suggest, 15.8 billion years old, our universe evolves as a coherent whole.

The tuning of the constants. Perhaps the most remarkable evidence for the coherence of the cosmos is the observed "fine-tuning" of its physical constants. The basic parameters of the universe have precisely the value that allows complex structures to arise. The fine-tuning in question involves upward of thirty factors and considerable accuracy. For example, if the expansion rate of the early universe had been one-billionth less than it was, the universe would have re-collapsed almost immediately; and if it had been one-billionth more, it would have flown apart so fast that it could produce only dilute, cold gases. A similarly small difference in the strength of the electromagnetic field relative to the gravitational field would have prevented the existence of hot and stable stars like the Sun, and hence the evolution of life on planets associated with these stars. Moreover, if the difference between the mass of the neutron and the proton were not precisely twice the mass of the electron, no substantial chemical reactions could take place, and if the electric charge of electrons and protons did not balance precisely, all configurations of matter would be unstable and the universe would consist of nothing more than radiation and a relatively uniform mixture of gases.

But even the astonishingly precisely adjusted laws and constants do not fully explain how the universe would have evolved out of the primordial radiation field. Galaxies had formed out of this radiation field when the expanding universe's temperature dropped to 3,000 degrees on the Kelvin scale. At that point the existing protons and electrons formed atoms of hydrogen, and these atoms condensed under gravitational pull, producing stellar structures and the giant swirls that make for the birth of galaxies. Calculations indicate that a very large number of atoms would have had to come together to start the formation of galaxies, perhaps of the order of 10^{16} suns. It is by no means clear how this enormous quantity of atoms—equivalent to the mass of

100,000 galaxies—would have come together. Random fluctuations among individual atoms do not furnish a plausible explanation.

A universe such as ours—with galaxies and stars, and life on this and presumably other life-supporting planets—is not likely to have come about as a matter of serendipity. According to Roger Penrose's calculations, the probability of hitting on our universe by a random selection from among the alternative-universe possibilities is one in $10^{10^{123}}$. This is an inconceivably large number, indicating an improbability of astronomical dimensions. Indeed, Penrose himself speaks of the birth of our universe as a "singularity" where the laws of physics do not hold.

Not even the surplus of matter over antimatter is explicable as a matter of pure chance: a randomly originating universe is not likely to have significantly violated the equivalence of charge and parity at its birth. That there is something (that is, some definite and observable "thing") rather than (nearly) nothing (no "thing") is not due merely to chance. Like the overall coherence of the cosmos, it is due to the presence of an active and effective kind of in-formation in nature.

COHERENCE IN THE LIVING WORLD

Quantum-type coherence. The coherence of the organism is quintessentially pluralistic and diverse at every level, from the tens of thousands of genes and hundreds of thousands of proteins and other macromolecules that make up a cell, to the many kinds of cells that constitute tissues and organs. The adjustments, responses, and changes required for the maintenance of the organism propagate in all directions simultaneously.

Quasi-instant, system-wide correlation cannot be produced solely by physical or even chemical interactions among molecules, genes, cells, and organs. Though some biochemical signaling—for example, of control genes—is remarkably efficient, the speed with which activating processes spread in the body, as well as the complexity of these processes, makes reliance on biochemistry alone insufficient. The conduction of signals through the nervous system, for example, cannot proceed faster than about twenty meters per second, and it cannot carry a large number of diverse signals at the same time. Yet there are quasi-instant, nonlinear, heterogeneous, and multidimensional correlations among all parts of the organism.

The level of coherence in the organism suggests that in some respects it is a macroscopic quantum system. Living tissue is a "Bose-Einstein condensate": a form of matter in which quantum-type processes—hitherto believed to be limited to the microscopic domain—occur at macroscopic scales. That they do has been verified in 1995, in experiments for which the physicists Eric A. Cornell, Wolfgang Ketterle, and Carl E. Wieman received the 2001 Nobel Prize. The experiments show that under certain conditions, seemingly separate particles and atoms interpenetrate as waves. For example, rubidium and sodium atoms behave not as classical particles but as nonlocal quantum waves, penetrating throughout the given condensate and forming interference patterns.

The quasi-instant connections that occur throughout the organism suggest that distant molecules and molecular assemblies resonate at the same or compatible frequencies. Whether the force that appears among the assemblies is attractive or repulsive depends on the given phase relations. For cohesion to occur among the assemblies, they have to resonate in phase—the same wave function must apply to them. This provision applies also to the coupling of frequencies among the assemblies. If faster and slower reactions are to accommodate themselves within a coherent overall process, the respective wave functions must coincide. They do in fact coincide, and as a consequence quantum biologists speak of the "macroscopic wave function" of the organism—a mathematical concept that gives formal expression to the instant connection that comes to light among all parts of the organism.

Hans-Peter Dürr, head of Germany's Max Planck Institute of

Physics, suggested an explanation of the coherence of living organisms in reference to the electromagnetic radiation that surrounds electrons in biomolecules. Consisting of billions of atoms, biomolecules resonate at frequencies between 100 and 1,000 gigahertz. Their longitudinal oscillations are linked to periodic charge displacements, giving rise to the radiation of electromagnetic waves of the same frequency. Dürr speculated that such specifically modulated carrier waves could interlink biomolecules, cells, and even entire organisms, whether they are contiguous or at considerable distance from each other.

Dürr concluded that—since according to quantum physics everything is included and incorporated in one indivisible potential reality—it should be possible to find many kinds of connecting links among phenomena. It is possible, he added, that some of these links may have less the character of a transmission of information between separate things that vibrate at the same frequency than the character of a genuinely nonlocal "communion" among seemingly separate but in reality deeply entangled particles and atoms, and the cells and molecules constituted of them.

The evolution of complex organisms. The historical fact that complex organisms have evolved on this planet is another indication of a hitherto unexplained form of coherence in the living world. It is evidence that the separation proposed by Darwin—between the genetic information encoded in the DNA of the cells of the organism and the phenome that results from this information—is not absolute. The genome does not mutate randomly, unaffected by the vicissitudes that befall the organism.

The idea that random mutations and natural selection are the basic mechanism of evolution was introduced in 1859, a full century before the nature of the hereditary material would be elucidated together with the specific mechanism by which heritable traits are transmitted. The identification of genes made up of strands of DNA came still later, followed by the discovery of the various modalities of mutation and rearrangement in the genome. The structure of genes in multicellular organisms was clarified in the late 1970s, sufficient DNA sequences to enable the analysis of the origin of genes became available in the 1980s, and the mapping of entire genomes began in the 1990s. Nevertheless, the basic mechanism of evolution described by Darwin was maintained unchanged. The "synthetic theory," the modern version of Darwinism, still insists that randomly produced genetic mutations and the chance fit of the mutants to the milieu evolve one species into another by producing new genes and new developmental genetic pathways, coding new and viable organic structures, body parts, and organs.

However, random mutations are not likely to produce viable species. The "search space" of possible genetic rearrangements within the genome is so enormous that random processes would take incomparably longer to produce new species than the time that was available for evolution on this planet. The probabilities are made a great deal worse by the consideration that many organisms, and many organs within organisms, are "irreducibly complex." A system is irreducibly complex, said the biologist Michael Behe, if its parts are interrelated in such a way that removing even one part destroys the function of the whole system. To mutate an irreducibly complex system into another viable system, every part has to be kept in a functional relationship with every other part throughout the entire transformation. Missing but a single part at a single step leads to a dead end. This level of constant precision is entirely unlikely to be achieved by random piecemeal modifications of the genetic pool.

An isolated genome working through random mutations is unlikely to produce a mutant organism capable of surviving in its milieu because it is not enough for a mutation to produce one or a few positive changes in a species; it must produce the full set. The evolution of feathers, for example, does not produce a reptile that can fly: radical changes in musculature and bone structure are also required, along with a faster metabolism to power sustained flight. The development of the eye requires thousands of mutations, finely coordinated with one another. The probability of a single mutation producing positive results is almost nil: statistically only one mutation in twenty million is likely to be viable; each mutation by itself is likely to make the organism less rather than more fit than it was. And if it is less fit, sooner or later it is eliminated by natural selection.

Already in 1937, the biologist Theodosius Dobzhansky noted that the sudden origin of a new species by random mutations might be an impossibility in practice. "Races of a species, and to a still greater extent species of a genus," he wrote, "differ from each other in many genes, and usually also in the chromosome structure. A mutation that would catapult a new species into being must, therefore, involve simultaneous changes in many gene loci, and in addition some chromosomal reconstruction. With the known mutation rates the probability of such an event is negligible." Dobzhansky did not give up the Darwinian theory; instead, he assumed that species formation is a slow and gradual process, occurring on a "quasi-geological scale."

The assumption of slow and gradual evolution was contradicted in the 1970s by the finding of new fossils: these show that the "missing links" that appear in the fossil record are not due to the incompleteness of the record, but are true jumps in the course of evolution. New species do not arise through a stepwise modification of existing species—they appear almost all at once. This finding prompted Stephen Jay Gould of Harvard and Niles Eldredge of the American Museum of Natural History to advance the theory of "punctuated equilibrium." In this macroevolutionary theory, new species arise in a time span of no more than five to ten thousand years. This may seem like a long time to humans, but as Gould and Eldredge pointed out, it translates into geological time as an instant.

The evidence is indirect but crystal clear: genetic mutations are far more successful in nature than random mutations are likely to be. If the genome is not guided by divine will or transcendent agency, it must be guided by the links of the integrated organism with its milieu. Thus we must conclude that not only are the parts of the organism nonlocally coherent; also the entire organism is nonlocally coherent with its larger environment.

COHERENCE IN CONSCIOUSNESS

Transpersonal connections. Linkages of a nonlocal kind obtain also in the world of consciousness. Regardless of separation in space and time, the consciousness of one person can be subtly linked with the consciousness of another.

So-called primitive peoples have long known of such "transpersonal" links. Medicine men and shamans appear capable of inducing telepathy through solitude, concentration, fasting, chanting, dancing, drumming, or psychedelic herbs. Whole clans seem able to remain in touch with each other no matter where their members roam. Australian Aborigines, the anthropologist A. P. Elkin found, are informed of the fate of family and friends even when they are beyond the range of sensory communication with them. A man far from his homeland will announce that his father is dead, or that his wife has given birth, or that there is some trouble in his country. He is so sure of his facts that he is ready to return home at once.

Tribal peoples, another anthropologist Marlo Morgan noted, are not only able to act on information received through their links with the consciousness of other people; they can also receive, and act on, information about certain aspects of their environment. Morgan noted that they are able to receive input from their environment, do something unique in decoding it, and then consciously act almost as if they had developed some tiny celestial receiver through which they receive universal messages. This observation was dramatically corroborated during the Asian tsunami catastrophe of December 2004. Sentinelese and other traditional tribes numbering merely in the hundreds live in the remote Andaman Islands in the Indian Ocean. They have been practically isolated from the rest of the world for thirty- to sixty-thousand years. It was expected that the tsunami would have taken a heavy toll among them, reducing some populations nearly to extinction. But this turned out not to be so: the tribespeople left for the highlands in time to escape the murderous waves. Some journalists speculated that they were informed of the coming danger by observing the behavior of animals. But this may not have been necessary: the tribespeople are likely to have preserved the kind of sensitivity to the environment that animals have. They could have accessed the signs of impending danger much as the birds and the elephants did.

Modern people seem to have lost access to this "celestial receiver," but laboratory experiments show that they have not lost the receiver itself. Under the right conditions, most people can become aware of the vague but meaningful images, intuitions, and feelings that come to them from other people and from their environment.

Transpersonal connections of this kind have been reported by a number of psychology and parapsychology laboratories. Thought and image-transference experiments between sender and receiver have been carried out involving distances ranging from half a mile to several thousand miles. Regardless of where they have been carried out and by whom, the success rate has been considerably above random probability. The receivers usually report a preliminary impression as a gentle and fleeting form. This form gradually evolves into a more integrated image. The image itself is experienced as a surprise, both because it is clear and because it is clearly elsewhere.

Such telepathic abilities may be widespread in the animal kingdom. Noted chimpanzee researcher Jane Goodall recounted that a female chimpanzee who had been particularly attached to her while she lived in the jungle, always showed up at her camp in Kenya when Goodall came to visit again. She would arrive on her own the day before Goodall herself did. Biologist Rupert Sheldrake carried out a series of surveys and experiments that indicate that household pets that have a close emotional bond with their owners are also informed of their owner's intentions even without sensory cues—they can in a sense "read their mind." In repeated surveys in

England and America more than half of dog owners and over one third of cat owners said that their pets were sometimes telepathic with them—knowing when they or a member of the household was on the way home or intended to go out, and responding at times to mere thoughts or silent commands.

Beyond telepathy, a related human ability is to synchronize the electrical activity of one's brain with the brain of others without sensory contact or communication. A series of experiments conducted by the Italian physician and brain researcher Dr. Nitamo Montecucco and witnessed by this writer showed that in deep meditation, the left and right hemispheres of the brain manifest identical wave patterns. Still more remarkably, the left and right hemispheres of different subjects become spontaneously synchronized. In one test, eleven of twelve mutually isolated meditators achieved a remarkable 98 percent synchronization of the full spectrum of their EEG waves.

An experiment carried out in the presence of this writer took place in southern Germany in the spring of 2001. At a seminar attended by about a hundred people, Dr. Günter Haffelder, head of the Institute for Communication and Brain Research of Stuttgart, measured the EEG patterns of Dr. Maria Sági, a trained psychologist and gifted natural healer, together with that of a young man who volunteered from among the participants. The young man remained in the seminar hall while the healer was taken to a separate room. Both the healer and the young man were wired with electrodes, and their EEG patterns were projected on a large screen in the hall. The healer diagnosed the health problems of the subject, while he sat with closed eyes in a light meditative state. When the healer found the subject's areas of organic dysfunction, she sent information designed to compensate for it. During the approximately fifteen minutes that the healer was concentrating on her task, her EEG waves dipped into the deep Delta region (between 0 and 3 Hz per second), with a few sudden eruptions of wave amplitude. This was surprising in itself, because when someone's brain waves descend into the Delta region, he or she is usually in state of deep sleep. But the healer was fully awake, in a state of intense concentration. Even more surprising was that the test subject exhibited the same Delta-wave pattern—it showed up in his EEG display about two seconds after it appeared in the EEG of the healer. Yet they had no sensory contact with each other.

Transcultural connections. Anthropological as well as laboratory evidence speaks to the reality of transpersonal connection between individuals, and archaeological and historical evidence testifies in turn that such connections exist also between entire cultures.

Subtle, spontaneous contact among cultures appears to have been widespread, as evidenced by the artifacts of different civilizations. In widely different locations and at different historical times, ancient cultures developed an array of similar artifacts and buildings. Although each culture added its own embellishments, Aztecs and Etruscans, Zulus and Malays, classical Indians and ancient Chinese built their monuments and fashioned their tools as if following a shared pattern. Giant pyramids were built in ancient Egypt as well as in pre-Columbian America, with remarkable agreement in design. The Acheulian hand ax, a widespread tool of the Stone Age, had a typical almond or tear-shaped design chipped into symmetry on both sides. In Europe this ax was made of flint, in the Middle East of chert, and in Africa of quartzite, shale, or diabase. Its basic form was functional, yet the agreement in the details of its execution in virtually all traditional cultures cannot be explained by the simultaneous discovery of utilitarian solutions to a shared need: trial and error is not likely to have produced such similarity of detail in so many far-flung populations.

Crafts, such as pottery making, took much the same form in all cultures. At this writer's suggestion, the University of Bologna historian Ignazio Masulli made an in-depth study of the pots, urns, and other artifacts produced by indigenous and independently evolving cultures in Europe, as well as in Egypt, Persia, India, and China during the period from the fifth to the second millennia B.C.E. Masulli found striking recurrences in the basic forms and designs but could not come up with a conventional explanation for them. The civilizations lived far apart in space and sometimes also in time, and did not seem to have had conventional forms of contact with each other.

Telesomatic connections. Transpersonal and transcultural phenomena are not limited to contact and communication between the consciousness of different people: repeatable and measurable effects can be transmitted also from the consciousness of one person to the *body* of another.

At the University of Nevada, the experimental parapsychologist Dean Radin undertook an experiment in which the test subjects created a small doll in their own image and provided various objects (pictures, jewelry, an autobiography, and personally meaningful tokens) to "represent" them. They also gave a list of what makes them feel nurtured and comfortable. These items and the accompanying information were used by the "healer"—who functioned analogously to the "sender" in thought- and image-transfer experiments—to create a sympathetic connection to the "patient." The latter was wired up to monitor the activity of his or her autonomous nervous system (electrodermal activity, heart rate, and blood pulse volume) while the healer was in an acoustically and electromagnetically shielded room in an adjacent building. The healer placed the doll and other small objects on the table in front of him and concentrated on them while sending randomly sequenced "nurturing" (active healing) and "rest" messages to the patient.

The electrodermal activity of the patients, together with their heart rate, was significantly different during the active nurturing periods than during the rest periods, while blood pulse volume was significant for a few seconds during the nurturing period. Both heart rate and blood flow indicated a "relaxation response"—which makes sense since the healer was attempting to "nurture" the subject via the doll. On the other hand, a higher rate of electrodermal activity showed that the patients' autonomic nervous system was becoming aroused. Why this should be so was puzzling, until the experimenters realized that the healers nurtured the patients by rubbing the shoulders or stroking the hair and face of the dolls that represented them. This, apparently, had the effect of a "remote massage" on the skin of the patients.

Radin and colleagues concluded that the local actions and thoughts of the healer are mimicked in the distant patient almost as if healer and patient were next to each other. Distance between sender and receiver seems to make no difference. This was confirmed in a large number of trials conducted by the experimental parapsychologists William Braud and Marilyn Schlitz regarding the impact of the mental imagery of senders on the physiology of receivers. Braud and Schlitz found that the mental images of the sender could reach out over space to cause changes in the physiology of the distant receiver. The effects are comparable to those that one's own mental processes produce on one's body. "Telesomatic" action by a distant person is similar to, and nearly as effective as, "psychosomatic" action by individuals on themselves.

Distant mental effect can be produced on other forms of life as well. In a series of experiments, the lie-detector expert Cleve Backster attached the electrodes of his lie detector to the leaves of a plant in his New York office. He recorded the changes in electric potentials on the surface of a leaf just as he would record such changes in a human subject. To his amazement, Backster found that the plant replicated his own emotions—showing sudden jumps and wild fluctuations at the precise moment when Backster himself had a strong emotional reaction, whether he was in the office or away from it. Somehow, the plant seemed to "read" his mind. Backster speculated that plants have a "primary perception" of the people and events around them.

Subsequently Backster tested many varieties of plants, cells, and even animals and found the same kind of response in the lie detector. The leaves of plants responded even when they were ground up and the remains distributed over the surface of the electrodes.

Backster then undertook a series of experiments in which he tested leukocytes (white cells) taken from the mouths of his test subjects. The procedure of obtaining the cells has been perfected for purposes of dentistry and produces a pure cell culture in a test tube. Backster moved the culture to a distant location, anywhere from five meters to twelve kilometers from his subjects. He placed the electrodes of the lie detector on the distant culture and provoked some emotion-producing response in his subjects. In one case he had a young man look at an issue of *Playboy* magazine. Nothing spectacular occurred until the young man came to the centerfold and saw a photo of actress Bo Derek in the nude. At that moment the needle of the lie detector attached to the distant cell culture began to swing, and kept fluctuating as long as the subject was looking at the picture. When he closed the magazine, the needle returned to trace a normal pattern, but was suddenly reactivated when the young man decided to have another look at the magazine.

In another test a former Navy gunner who was at Pearl Harbor during the Japanese attack was shown a TV program depicting the attack. He showed no particular reaction until the face of a Navy gunner appeared on the screen, followed by a shot of a Japanese plane falling into the sea. At that moment the needle of the lie detector attached to his cells twelve kilometers away jumped. Subsequently, both he and the young man with the *Playboy* magazine confirmed that they had had strong emotional reactions at these particular points in the experiments.

It made no difference whether the cells were a few meters or several kilometers away. The lie detector displayed exactly the response it would have displayed if it had been attached directly to the subject's body. Backster concluded that a form of "biocommunication" is taking place for which there is no adequate explanation.

A WORD IN CONCLUSION

There *is* an explanation for the phenomena that puzzle today's frontline investigators; we *can* understand what processes underly the nonlocal coherence of the human body, of all life, of the quantum, and of the entire universe. *It is the presence of in-formation throughout the cosmos, carried and conveyed by the universal in-formation field we have named the Akashic field.* The action of this subtle but real A-field explains the nonlocality of the smallest measurable units of the universe as well as of its largest observable structures. It explains the coherence of living organisms, and their coherence with the milieu in which they live and evolve. It also explains the coherence of the human brain, and of the consciousness associated with it, in regard to the brain and consciousness of other human beings, and even the world at large. And, last but not least, it explains the astounding fact that the physical parameters of the universe are so finely adjusted that living organism can exist and evolve on this planet, and possibly on countless other planets in this and in other galaxies.

There is no need to ascribe nonlocal coherence, the remarkable space- and time-transcending connection of everything with everything else, to the action of divine will, or to forces above or beyond the natural world. Nonlocal coherence is a bona fide scientific phenomenon, just as real and understandable as light, electromagnetism, mass, and gravitation—although initially they prove to be just as puzzling as they have been. A-field in-formation is the logical explanation of nonlocal coherence: of the mysterious way in which quanta are connected across space and time, of the evident but nonetheless astounding fact that we and other organisms have evolved and can live on this planet and, last but not least, of the seemingly miraculous capacity of the universe to bring forth human beings such as you and I, who now ask themselves why this universe is so well tuned that in all essential respects it is both instantly and universally interconnected.

Over Four Decades in Quest
of an Integral Theory
# of Everything

An Autobiographical Retrospective

Science and the Akashic Field is the product of more than four decades of searching for meaning through science. I started on this quest in the spring of 1959, shortly after my first son was born. Until then my interest in philosophical and scientific questions had been just a hobby—I had been traveling the world as a musician, and nobody, not even I, had ever suspected that it would become more than an intellectual pastime. But my interest in finding a meaningful and encompassing answer to what I experienced and knew about life and the universe grew, and the quest that began in 1959 became an all-consuming vocation. It culminated four decades later in the spring of 2001, as I sat down to draft out *The Connectivity Hypothesis*, my latest theoretical work. The original edition of the present book, summarizing my findings for the general readership, followed in 2002–2004. It is here updated with the latest findings from the sciences, and my latest, most mature interpretation of the findings.

My enduring interest has been to find an answer to questions such as "What is the nature of the world?" and "What is the meaning of my life in the world?" These are typically philosophical questions—although the majority of today's academic philosophers prefer to hand them to theologians and poets—yet I did not seek to answer them through theoretical philosophy. While I was not an experimental scientist (and given my background and interest I was not attempting to become one), I did have a strong sense that the best way to tackle these questions is through science. Why? Simply because empirical science is the human endeavor that is the most rigorously and systematically oriented toward finding the truth about the world, and testing its findings against observation and experience. I wanted the most reliable kind of answers there are, and reflected that I could find no better source for them than science.

For a young man in his twenties without formal background in a specific field of science, this was quite presumptuous. I would like to call what I had intellectual courage, but at the time I did not feel especially courageous—just curious and committed. Nonetheless, I was not entirely unprepared, for I had done a good deal of prior reading (mostly on planes and trains and in hotel rooms) and took part in various college and university courses. Being a successful concert pianist, I never enrolled for an academic degree for which I saw no conceivable use.

In 1959 I turned over a new leaf: I set about doing systematic reading and research. What was until then a favorite hobby became a methodical quest. I started with the foundations of science in classical Greek thought and moved to the founders of modern science before turning to contemporary science. I was interested neither in the technical details that take up the lion's share of the training of science professionals—techniques of research, observation, and experimentation—nor in controversies about methodological or historical fine points. I wanted to get straight to the heart of the matter: to find out what a given science could tell me about the segment of nature it investigates. This required a good deal of spadework. The findings were unexpectedly sparse, consisting of a few concepts and statements, usually at the end of extensive mathematical and methodological treatises. They were, however, extremely valuable, much like nuggets of gold that come to hand after sifting through streams of water and mountains of ore.

In the course of the 1960s, I learned to do my sifting rapidly and efficiently, covering a good deal of ground. What meaning I found half-buried in particular fields I jotted down, and attempted to bring it into relation with what I found in other fields. I did not intend to write a treatise or create a theory, I just wanted to understand what the world and life—my life, and life in general—are all about. I made copious notes, but never expected that they would get into print. How they did so is one of the curious episodes of my life.

After a successful concert in The Hague, I found myself sitting at late supper next to a Dutchman who brought up some of the very questions that fascinated me. I got into conversation with him, and ended by going up to my hotel room to show him the notes I always had with me. He retired into a corner and began reading. Shortly after that he disappeared. I was concerned, since I had no copy. However, the next morning my newfound friend reappeared with my notes under his arm. He announced that he wanted to publish them. This was a surprise, for I knew neither that he was a publisher (he turned out to be the philosophy editor at the renowned Dutch publishing house Martinus Nijhoff) nor that my notes would merit publication. Of course, they required a good deal of completing and organizing before they could be published in book form. But published they were, a year and a half later (*Essential Society: An Ontological Reconstruction*, 1963).

The experience in The Hague reinforced my determination to pursue my quest. I joined the Institute of East European Studies at Switzerland's University of Fribourg, and for several years combined writing and research with concert work. I came out with another, less theoretical, book shortly after the first (*Individualism, Collectivism, and Political Power*, 1963) and a few years later published another philosophical treatise (*Beyond Scepticism and Realism*, 1966). The period of writing and researching combined with concertizing came to an end when, in 1966, I received an invitation from Yale University's Department of Philosophy to spend a semester there as visiting fellow. Accepting that invitation was a major decision, for it meant exchanging the concert stage for the life of an academic.

The decision to go to Yale—which led to teaching appointments at various U.S. universities and, in 1969, to a Ph.D. at the Sorbonne in Paris—gave me the opportunity to pursue my quest full time. Although in any established university there is considerable pressure to keep to the rather narrowly defined territory of one's own field, I never wavered from the conviction that there is meaning to be discovered in the world at large, and that the best way of discovering it is to query the theories put forward by leading scientists in all the relevant fields, not just those that belong to one's area of specialization. I was fortunate in finding colleagues—first at Yale, then at the State University of New York—who understood this conviction and helped me overcome the academic hurdles that would have stood in the way.

The search for meaning through science called for considerable time and energy. I soon realized that, like Archimedes, I needed firm ground from which to start. I found two basic alternatives. One was to start with the stream of one's own conscious experience and see what kind of world one could logically derive from that experience. The other was to gather all the information one can about the world at large, and then see if one can account for one's own experience as the experience of that world. The former has been the method of the empirical schools of Anglo-Saxon philosophy and of that branch of continental philosophy that took its cue from Descartes, and the latter the method of naturalistic metaphysics and science-based philosophy. I read up on these schools, paying special attention to Bertrand Russell and Alfred Ayer among the British philosophers, Edmund Husserl and the phenomenologists of the continental schools, and Henri Bergson and Alfred North Whitehead among the naturalistic process philosophers. I concluded that neither the formal analysis of experience nor the introspective method of the phenomenologists leads to a meaningful concept of the real world. These schools ultimately get bogged down in what philosophers call the "ego-centric predicament." It appears that the more systematically one investigates one's immediate experience, the less easy it is to get beyond it to the world to which that experience presumably refers. We are logically obliged to take the initial leap of assuming the objective existence of the external world, and then to create the scheme in light of which our experience makes sense as the human experience of that world.

In *Beyond Scepticism and Realism,* I contrasted the "inferential" approach that starts from one's own experience with the alternative "hypothetico-deductive" method that envisages the nature of the world and explores how our observations accord with it. I concluded that, ideally, the overlap between these distinct and sometimes seemingly contradictory approaches is what gives the most reliable information about the real nature of the world. I identified some areas of overlap, but did not stop there: I wanted to get on with my quest, and began to explore the bold hypothetico-deductive approach. To my considerable relief, I found that this approach had been adopted by many great philosophers and practically all theoretical scientists, from Newton and Leibniz to Einstein and Eddington.

Einstein stated the principal premise of the naturalistic approach. "We are seeking," he said, "for the simplest possible scheme of thought that will bind together the observed facts." The simplest possible scheme, I realized, cannot be inferred from observations: as Einstein said, it needs to be imaginatively envisaged. One must search for and codify the relevant observations, but one cannot stop there. While empirical research is necessary, the creative task of putting together the resulting data in ways that they make sense as meaningful elements of a coherent system cannot be neglected. It is the principal challenge facing the inquiring mind. The attempt to "create the simplest possible scheme of thought that will bind together the observed facts" (and by "observed facts" I meant all the facts needed to make sense of the world) defined my intellectual agenda for the next four decades.

The scheme I first envisaged rested on the organic metaphysics of Whitehead. In this conception, which dated originally from the 1920s, the world and all things in it are integrated and interacting "actual entities" and "societies of actual entities." Reality is fundamentally organic, so living organisms are but one variety of the organic unity that emerges in the domains of nature. My subsequent readings in cosmology and biology confirmed the soundness of this assumption. Life, and the cosmos as a whole, evolves as integrated entities within a network of constant formative interaction. Each thing not only "is," it also "becomes." Reality, to cite Whitehead, is process, and an integrative evolutionary process at that.

The question I asked was how I could identify the evolving entities of the world in such a way that they would make sense as elements in an organically integral universe. Colleagues at Yale called my attention to the work of Ludwig von Bertalanffy in the area of "general system theory." Bertalanffy was attempting to integrate the field of biology in an overall scheme that would lend itself to further integration with other domains of natural science, and even with the human and social sciences. His key concept was "system," conceived as a basic entity in the world. Systems, he argued, appear in similar ("isomorphic") ways in physical nature, living nature, as well as the human world. This was most helpful: it supplied the conceptual tool I was looking for. I read Bertalanffy, then met with him and developed the concept of what we jointly decided to call "systems philosophy."

Introduction to Systems Philosophy (1972) was a painstakingly researched book—it took five years to write—and when it was published I was tempted to rest for a while on my laurels. But I was not satisfied. I needed to find an answer in leading-edge science not only to how systems are constituted and how they relate to each other, but also to how they change and evolve. Whitehead's metaphysics gave me the general principles and Bertalanffy's general system theory clarified the relations between systems and environments. What I still needed was the key to understanding how these relations can lead to integrative and on the whole irreversible evolution in the biosphere, and in the universe as a whole.

To my surprise, the key was furnished by a discipline about which I knew little at the time: nonequilibrium thermodynamics. I reached this conclusion on the basis of my brief but intense friendship with Erich Jantsch, who died unexpectedly a few years later. He directed my attention to the work, and subsequently to the person, of the Russian-born Nobel laureate thermodynamicist Ilya Prigogine. The latter's concept of "dissipative structures" that are subject to periodic "bifurcations" furnished the evolutionary dynamic I needed. After discussing this concept with Prigogine, my work focused on what I called "general evolution theory." The basic kind of entity that populates the world transformed in my thinking from Whitehead's "organism" and Bertalanffy's "general system" to Prigogine's nonlinearly bifurcating "dissipative structure," an evolving thermodynamically open system. The world began to make more and more sense.

Apparently, the sense I made of the world also intrigued scholars in fields other than systems theory and philosophy. While teaching and researching at the State University of New York at Geneseo, to my surprise I received a phone call from Richard Falk, of Princeton University's Center of International Studies. Falk, one of the foremost "world system" theorists of the time, asked me to come to Princeton to lead a series of seminars on the application of my systems theory to the study of the international system. I assured him that I knew next to nothing about the international system and had only vague notions of how my theory would apply to it. But Falk was not to be deterred. He and his colleagues, he said, would see to the application of my theory if I would come and discuss that theory with them. This I agreed to do.

The experience of my Princeton seminars was intellectually rewarding as well as exciting: it opened new vistas. I found a new and intensely practical application for general system theory, systems philosophy, and general evolution theory: human society and civilization. Society and civilization, I realized in the mid-1970s, are undergoing a process of irreversible transformation. The human world is growing beyond the bounds of the nation-state system to the limits of the globe and the biosphere. This called for rethinking some of our most cherished notions about how societies are structured, how they operate, and how they develop. With valuable input from Richard Falk and other Princeton colleagues, I spelled out my evolutionary conception of the world system in *A Strategy for the Future: The Systems Approach to World Order* (1974).

Strategy elicited attention beyond academia. Another call followed, this time from Aurelio Peccei, the visionary Italian industrialist who founded the world-renowned think tank known as the Club of Rome. He suggested that I apply the systems approach to the "limits to growth" problem, focusing not on the limits themselves (as Jay Forrester and Dennis and Donella Meadows did in the first report to the club, *The Limits to Growth*), but on the ambitions and motivations that drive people and societies to encounter the limits. This invitation was an intellectual challenge with major practical relevance—it could not be refused. I took a leave of absence from my university and moved to the United Nations headquarters in New York. Davidson Nicol, executive director of the UN's Institute of Training and Research (UNITAR), invited me to join his institute in order to create the international team that was to work on this project. Within a year, some 130 investigators on six continents were enlisted in creating the Club of Rome's third report, focusing on humankind's "inner" rather than "outer" limits (*Goals for Mankind: The New Horizons of Global Community,* 1977).

Having finished the report, I repaired to my university to resume researching, writing, and teaching. This, however, was not to be. A further call from Nicol asked me to represent UNITAR at the founding of the United Nations University in Tokyo, and when I filed my report Nicol asked me to stay on at the institute to head research on the hottest subject of the day, the "new international economic order." This was another challenge that could not be ignored. After three years of intense work, fifteen volumes written with collaborators from ninety research institutes in every part of the world were published in a series created for this purpose by Pergamon Press of Oxford: the New International Economic Order Library. The NIEO Library was to produce background documentation for the General Assembly's landmark General Session of 1980, which was to launch the "global dialogue" between the developing South and the industrialized North. But the big powers of the North refused to enter the dialogue and the UN system dropped the project of the new international economic order.

When I was about to return to my university to pursue at last my principal quest, UN Secretary-General Kurt Waldheim asked me to suggest other ways in which North–South cooperation could be pursued. The proposal I made to him and to UNITAR was based on systems theory: it was to insert another "systems level" between the level of individual states and the level of the United Nations. This was the level of regional social and economic groupings. The project, called Regional and Interregional Cooperation, was adopted by UNITAR and took four years of intense work to carry out. In 1984 I reported the results in four bulky volumes that accompanied a declaration of a specially convened "panel of eminent persons." Due to internal politics, the declaration was not handed to the secretary-general and thus could not be made into an official document, but its text was circulated to all member-state delegations. Disappointed with this outcome but hopeful that sooner or later the proposals contained in the declaration would bear fruit, I decided that I had merited a sabbatical year. I moved with my family to our converted farmhouse in Tuscany. That sabbatical year, begun in 1984, has not yet come to an end.

However, the 1980s and '90s turned out to be much more than

a "read and write" sabbatical. It was a time of increasingly intense international commitments. In the 1980s I was involved in discussions at the Club of Rome, then took a major part in the United Nations University's European Perspectives project. Subsequently, I served as science adviser to Federico Mayor, the two-term director-general of UNESCO. But since 1993 the brunt of my attention was focused on the Club of Budapest, the international think-tank I founded that year to do what I had hoped the Club of Rome would do: center attention on the evolution of human values and consciousness as the crucial factors in changing course—from a race toward degradation, polarization, and disaster to a rethinking of values and priorities so as to navigate today's transformation in the direction of humanism, ethics, and global sustainability. As reports to the Club of Budapest I wrote *Third Millennium: The Challenge and the Vision* (1997) and most recently *You Can Change the World: The Global Citizen's Handbook for Living on Planet Earth* (2003).

Notwithstanding these activities and commitments, I remained faithful to my basic quest. When in 1984 I left the United Nations for the Tuscan hills, I took stock of how far I had gotten. I found that I needed to go further. Systems theory, even with the Prigoginian dynamic, provided a sophisticated but basically local explanation of how things relate and evolve in the world. The open system dynamic of evolution refers to particular systems; their interaction with other systems and the environment constitutes what Whitehead termed "external" relations. Yet Whitehead affirmed that in the real world all relations are *internal*: every "actual entity" is what it is because of its relations to all other actual entities. With this in mind, I set about reviewing the latest findings in quantum physics, evolutionary biology, cosmology, and consciousness research, and found that the idea of internal relations is entirely sound. Things in the real world are indeed strongly—"internally," "intrinsically," and even "nonlocally"— connected and correlated with each other.

Internal relations also bind our own consciousness with the

consciousness of others. This was brought home to me in 1986 by a personal experience that I recounted in 1993 in the preface to *Creative Cosmos* and will not repeat here. Although a mystical experience does not provide proof of internal relations between one's mind and the mind of others, it does provide an incentive to study the possibility that such relations exist. This consideration became part of my explorations in the years that followed. They were reinforced by my experience of the remarkable faculties of social psychologist and Club of Budapest colleague Dr. Mária Sági. For a period of over twenty years, she has consistently and correctly diagnosed whatever health problems I have experienced whether I was near her or at a distant location, and found the appropriate, remarkably effective homeopathic remedies to treat them.

In our discussions and experiments it became clear to me that in her healing work she is receiving reliable extrasensory information from a source that must be accounted for in any serious discourse about the nature of reality—and that this is likely to be the same source that creates entanglement among quanta and transpersonal connections among organisms and minds. Over the years I have named this source first the QVI (quantum/vacuum interaction) field, then the Ψ- (psi-) field, and presently the Akashic or A-field.

In the science books I have written in the "Tuscan period" (since the late 1980s) I marshal evidence that through this universal field all things are constantly and enduringly connected. In these books (which include, in addition to the one in the hands of the reader, *The Creative Cosmos, The Interconnected Universe, The Whispering Pond, The Connectivity Hypothesis,* and *Science and the Reenchantment of the Cosmos*) I present experimental evidence for this field, as well as a progressively elaborated theoretical explanation of it. I put forward a science-based scheme for binding together the remarkable facts of correlation, connection, and coherence that come to light at the cutting edges of the physical and biological sciences, and in the emerging discipline of systematic consciousness research. Researching and developing such a scheme would be of the utmost importance for science as well as for society. It would bring us closer to Einstein's goal of finding the "simplest possible scheme that can bind together the observed facts"—and thereby lend scientifically grounded meaning to the whole of our experience, as well as to our place in the universe.

References

INTRODUCTION

Peat, F. David. *Synchronicity: The Bridge Between Matter and Mind.* New York: Bantam Books, 1987.

Tarnas, Richard. *Cosmos and Psyche: Intimations of a New World View.* New York: Ballantine, 2006.

Weinberg, Steven. "Lonely planet." *Science and Spirit* 10 (April–May 1999).

CHAPTER 1:
THE CHALLENGE OF AN INTEGRAL
THEORY OF EVERYTHING

Bohm, David. *Wholeness and the Implicate Order.* London: Routledge & Kegan Paul, 1980.

Smolin, Lee. *The Trouble with Physics: The Rise of String Theory, The Fall of Science, and What Comes Next.* New York: Houghton Mifflin, 2006.

Wilber, Ken. *A Theory of Everything: An Integral Vision for Business, Politics, Science and Spirituality.* Boston: Shambhala, 2000.

Woit, Peter. *Not Even Wrong: The Failure of String Theory and the Search for Unity in Physical Law.* New York: Basic Books, 2006.

CHAPTER 2:
ON PUZZLES AND FABLES

Bekenstein, Jacob D. "Information in the holographic universe." *Scientific American* (August 2003).

Everett, Hugh. "'Relative State' Formulation of Quantum Mechanics." *Rev. Mod. Physics* 29 (July 1957).

Gefter, Amanda. "Mr. Hawking's Flexiverse." *New Scientist* (22 April 2006).

CHAPTER 3:
A CONCISE CATALOG OF
THE PUZZLES OF COHERENCE

Aspect, A., and P. Grangier. "Experiments on Einstein-Podolsky-Rosen-type correlations with pairs of visible photons." In *Quantum Concepts in Space and Time*, R. Penrose and C. J. Isham, eds. Oxford: Clarendon Press, 1986.

Bischof, Marco. "Field concepts and the emergence of a holistic biophysics." In *Biophotonics and Coherent Structures*, L. V. Beloussov, V. L. Voeikov, and R. Van Wijk, eds. Moscow: Moscow University Press, 2000.

Bonanos, Alceste Z., et al. "The first DIRECT distance determination to a detached eclipsing bionary in M33." *The Astrophysical Journal* 652 (2006).

Bucher, Martin A., Alfred S. Goldhaber, and Neil Turok. "Open universe from inflation." *Physical Review D* 52, no. 6 (15 September 1995).

Byrd, R. C. "Positive therapeutic effects of intercessory prayer in a coronary care population." *Southern Medical Journal* 81, no. 7 (1988).

Dawkins, Richard. *The Blind Watchmaker*. London: Longmans, 1986.

Dossey, Larry. "Era III medicine: the next frontier." *ReVision* 14, no. 3 (1992).

Einstein, Albert, Boris Podolski, and Nathan Rosen. "Can quantum mechanical description of physical reality be considered complete?" *Physical Review* 47 (1935).

Frazer, Sir James G. *The Golden Bough: A Study in Magic and Religion*. 13 Vols. London: MacMillan, 1890.

Grinberg-Zylberbaum, Jacobo, M. Delaflor, M. E. Sanchez-Arellano, M. A. Guevara, and M. Perez. "Human communication and the electrophysiological activity of the brain." *Subtle Energies* 3, no. 3 (1993).

Guth, Alan H. *The Inflationary Universe: The Quest for a New Theory of Cosmic Origins*. New York: Perseus Books, 1997.

Hagley, E., et al. "Generation of Einstein-Podolsky-Rosen pairs of atoms." *Physical Review Letters* 79, no. 1 (1997): 1–5.

Harris, W. S., M. Gowda, J. W. Kolby, C. P. Strycharz, J. L. Varck, P. G. Jones, et al. "A randomized control trial of the effects of remote, intercessory prayer on outcomes in patients admitted to the coronary care unit." *Arch. Intern. Med.* 159 (1999).

Ho, Mae-Wan. *The Rainbow and the Worm: The Physics of Organisms*. Singapore and London: World Scientific, 1993.

Hoyle, Fred, G. Burbidge, and J. V. Narlikar. "A quasi-steady state cosmology model with creation of matter." *The Astrophysical Journal* 410 (20 June 1993).

Keen, Jeffrey S. "Mind-created dowsable fields." *Dowsing Research Group: The First 10 Years*. Wolverhampton, U.K.: Magdalena Press, 2003.

Lieber, Michael M. "Hypermutation as a means to globally restabilize the genome following environmental stress." *Mutation Research, Fundamental and Molecular Mechanisms of Mutagenesis* 421, no. 2 (1998).

Maniotis, A., et al. "Demonstration of mechanical connections between integrins, cytoskeletal filaments, and nucleoplasm that stabilize nuclear structure." *Proceedings of the National Academy of Sciences U.S.A.* 94, no. 3 (1997).

Playfair, Guy. *Twin Telepathy: The Psychic Connection*. London: Vega Books, 2002.

Prigogine, I., J. Geheniau, E. Gunzig, and P. Nardone. "Thermodynamics of Cosmological Matter Creation." *Proceedings of the National Academy of Sciences U.S.A.* 85 (1988).

Puthoff, Harold, and Russell Targ. "A perceptual channel for information transfer over kilometer distances: historical perspective and recent research." *Proceedings of the IEEE* 64 (1976).

Schäfer, Lothar. "Quantum Reality, the Emergence of Complex Order from Virtual States, and the Importance of Consciousness in the Universe." *Zygon* 41, no. 3 (September 2006).

Steinhardt, Paul J., and Neil Turok. "A cyclic model of the universe." *Science* 296 (2002).

Targ, Russell, and Harold A. Puthoff. "Information transmission under conditions of sensory shielding." *Nature* 251 (1974).

Tittel, W., J. Brendel, H. Zbinden, and N. Gisin. "Violation of Bell Inequalities by Photons More than 10 KM Apart." *Phys. Rev. Lett.* 81 (1998): 3563–3566.

CHAPTER 4:
THE CRUCIAL SCIENCE FABLE—
IN-FORMATION IN NATURE

Akimov, A. E., and G. I. Shipov. "Torsion fields and their experimental manifestations." *Journal of New Energy* 2, no. 2 (1997).

Akimov, A. E., and V. Ya. Tarasenko. "Models of polarized states of the physical vacuum and torsion fields." *Soviet Physics Journal* 35, no. 3 (1992).

Bell, John S. "On the Einstein-Podolsky-Rosen Paradox." *Physics* 1 (1964).

Clarke, Chris. "Entanglement—the explanation for everything?" *Network Review* 86 (Winter 2004).

Franks, Felix. Reported in: Robert Matthews, "The quantum elixir." *New Scientist* 8 (April 2006).

Gazdag, László. *Beyond the Theory of Relativity*. Budapest: Robottechnika Kft., 1998.

Haisch, Bernhard, Alfonso Rueda, and H. E. Puthoff. "Inertia as a zero-point-field Lorentz force." *Physical Review A* 49, no. 2 (1994).

Maxwell, James Clerk. *Treatise on Electricity and Magnetism*. Oxford: Clarendon Press, 1873.

Mitchell, Edgar R. *The Way of the Explorer: An Apollo Astronaut's Journey through the Material and Mystical Worlds.* New York: Putnam, 1996.

Puthoff, H. E. "Quantum vacuum energy research and 'metaphysical' processes in the physical world." *MISAHA Newsletter* 32–35 (January–December 2001).

Sakharov, A. "Vacuum quantum fluctuations in curved space and the theory of gravitation." *Soviet Physics-Doklady* 12, no. 11 (1968).

Shipov, G. I. *A Theory of the Physical Vacuum: A New Paradigm.* Moscow: International Institute for Theoretical and Applied Physics RANS, 1998.

Vivekananda, Swami. *Raja-Yoga.* Calcutta: Advaita Ashrama, 1982.

CHAPTER 5:
THE ORIGINS AND DESTINY OF LIFE
AND THE UNIVERSE

Dawkins, Richard. *The Theory of Evolution.* Cambridge, U.K.: Cambridge University Press, 1993.

Drake, Frank. *Intelligent Life in Space.* New York: Macmillan, 1964.

Huang, Su-Shu. "Occurence of Life in the Universe." *American Scientist* 47 (1959): 397–402.

Ponnamperuma, Cyril. "Experimental studies on the origin of life." *Journal of the British Interplanetary Society* 42 (1989).

———. "The origin, evolution, and distribution of life in the universe." In *Cosmic Beginnings and Human Ends*, Clifford N. Matthews and Roy A. Varghese, eds. Chicago and La Salle, Ill.: Open Court, 1995.

Sagan, Carl. *Intelligent Life in the Universe.* New York: Emerson Adams Press, 1966.

Shapley, Harlow. *Of Stars and Men.* Boston: Beacon, 1958.

Taormina, Robert J. "A New Consciousness for Global Peace." In *Proceedings, Third International Symposium on the Culture of Peace.* Baden Baden, Germany, 1999.

Ward, Peter B. *Rare Earth: Two Tiers of Life in the Universe.* New York: Springer Verlag, 2000.

CHAPTER 6:
CONSCIOUSNESS—HUMAN AND COSMIC

Aurobindo, Sri. *The Life Divine.* 2nd printing. New York: Sri Aurobindo Library, 1951.

Bache, Chris. Letter to the author, July 2005.

Bailey, Alice. *Telepathy and the Etheric Vehicle.* New York: Lucis, 1950.

Beck, Don, and Christopher C. Cowan. *Spiral Dynamics: Mastering Values, Leadership and Change.* Oxford, U.K.: Blackwell, 1996.

Botkin, Allan, and R. Craig Hogan. *Reconnections: The Induction of After-Death Communication in Clinical Practice.* Charlottesville, Va.: Hampton Roads, 2006.

Chalmers, David J. "The puzzle of conscious experience." *Scientific American* 273 (December 1995).

Dyson, Freeman. *Infinite in All Directions.* New York: Harper & Row, 1988.

Fechner, Gustav. Quoted in William James, *The Pluralistic Universe.* London, New York, and Bombay: Longmans, Green & Co., 1909.

Fodor, Jerry A. "The big idea." *New York Times Literary Supplement,* 3 July 1992.

Gebser, Jean. *Ursprung und Gegenwart.* Stuttgart: Deutsche Verlagsanstalt, 1949.

Greyson, B. "Incidence and correlates of near-death experiences in a cardiac care unit." *Gen. Hosp. Psychiatry* 25, no. 4 (July–August 2003).

Grof, Stanislav. *The Cosmic Game: Explorations at the Frontiers of Human Consciousness.* Albany: State University of New York Press, 1999.

Lommel, W. van. "About the Continuity of our Consciousness." In *Brain Death and Disorders of Consciousness,* C. Machado, and D. A. Shewmon, eds. New York, London: Kluwer Academic/Plenum Publishers, 2004.

———. "Near-death experience, consciousness, and the brain." *World Futures* 62, nos. 1–2 (2006).

Lommel, W. van, R. Wees, V. Meyers, and I. Elfferich. "Near-death experience in survivors of cardiac arrest: a prospective study in the Netherlands." *Lancet* 358 (2001).

Marcer, Peter. "The brain as a conscious system." *International Journal of General Systems* 27 (1998).

Marcer, Peter, and W. Schempp. "Model of the neuron working by quantum holography." *Informatica* 21 (1997).

Mitchell, Edgar R. *The Way of the Explorer: An Apollo Astronaut's Journey through the Material and Mystical Worlds.* New York: Putnam, 1996.

Parnia, Sam, and Peter Fenwick. "Near-death experiences in cardiac arrest: visions of a dying brain or visions of a new science of consciousness." *Resuscitation* 52 (2002): 5–11.

Stevenson, Ian. *Children Who Remember Previous Lives.* Charlottesville: University Press of Virginia, 1987.

———. *Cases of the Reincarnation Type.* 4 vols. Charlottesville: University Press of Virginia, 1975–83.

Wagenseil, Sabine. "Tod ist nicht tödlich: durchgaben über den Tod von einem Toten" [Death is not deadly: Transmissions about death from a dead]. *Grenzgebiete der Wisseenschaft* 51 (2002).

Whitehead, Alfred North. *Process and Reality.* Cambridge, U.K.: Cambridge University Press, 1929.

Wilber, Ken. *Up from Eden: A Transpersonal View of Human Evolution.* Boulder, Colo.: Shambhala, 1983.

THE PHENOMENON OF COHERENCE

Backster, Cleve, and Steve White. "Biocommunications capability: Human donors and in vitro leukocytes." *International Journal of Biosocial Research* 7, no. 2 (1985).

Behe, Michael J. *Darwin's Black Box: The Biochemical Challenge to Evolution.* New York: Touchstone Books, 1998.

Braud, W. G. and M. Schlitz. "Psychokinetic influence on electrodermal activity." *Journal of Parapsychology* 47 (1983).

Buks, E., R. Schuster, M. Heiblum, D. Mahalu, and V. Umansky. "Dephasing in electron interference by a 'which-path' detector." *Nature* 391 (26 February 1998).

Dobzhansky, Theodosius. *Genetics and the Origin of Species,* 2nd ed. New York: Columbia University Press, 1982.

Dürr, Hans-Peter. "Sheldrake's ideas from the perspective of modern physics." *Frontier Perspectives* 12 (Spring 2003).

Dürr, S., T. Nonn, and G. Rempe. "Origin of quantum-mechanical complementarity probed by a 'which-way' experiment in an atom interferometer." *Nature* 395 (3 September 1998).

Eldredge, Niles, and Stephen J. Gould. "Punctuated equilibria: an alternative to phylogenetic gradualism." In *Models in Paleobiology,* Thomas J. M. Schopf, ed. San Francisco: Freeman, Cooper, 1972.

Elkin, A. P. *The Australian Aborigines.* Sydney: Angus & Robertson, 1942.

Gould, Stephen J., and Niles Eldredge. "Punctuated equilibria: the tempo and mode of evolution reconsidered." *Paleobiology* 3 (1977).

Grof, Stanislav. *The Adventure of Self-discovery.* Albany: State University of New York Press, 1988.

———. *Psychology of the Future: Lessons from Modern Consciousness Research.* Albany: State University of New York Press, 2000.

Heisenberg, Werner. *Physics and Philosophy.* New York: Harper & Row, 1985.

Kafatos, Menas, and Robert Nadeau. *The Conscious Universe: Part and Whole in Modern Physical Theory.* New York: Springer Verlag, 1990, 1999.

Masulli, Ignazio. "Recurrences of form in the Old World as evidence of collective consciousness: a hypothesis for historical research." *World Futures* 48, nos. 1–4 (1997).

Montecucco, N. "Cyber: Ricerche Olistiche." *Cyber* (November 1992).

Morgan, Marlo. *Mutant Message Down Under.* New York: HarperCollins, 1991.

Nadeau, Robert, and Menas Kafatos. *The Non-Local Universe: The New Physics and Matters of the Mind.* New York: Oxford University Press, 1999.

Radin, Dean. *The Conscious Universe: The Scientific Truth of Psychic Phenomena.* San Francisco: HarperEdge, 1997.

Ring, Kenneth. "Near-death and out-of-body experiences in the blind: A study of apparent eyeless vision." *Journal of Near-Death Studies* 16 (Winter 1997).

Sági, Maria. "Healing through the QVI-field." In *The Evolutionary Outrider,* David Loye, ed. London: Adamantine Press, 1998.

Schrödinger, Erwin. *My View of the World.* Cambridge, U.K.: Cambridge University Press, 1964.

Sheldrake, Rupert, C. Lawlor, and J. Turney. "Perceptive pets: A survey in London." *Biology Forum* 91 (1998).

Wheeler, John A. "Bits, quanta, meaning." In *Problems of Theoretical Physics,* A. Giovannini, F. Mancini, and M. Marinaro, eds. Salerno, Italy: University of Salerno Press, 1984.

———. "Quantum cosmology." In *World Science,* L. Z. Fang and R. Ruffini, eds. Singapore: World Scientific, 1987.

Zou, X. Y., L. J. Wang, and L. Mandel. "Induced coherence and indistinguishability in optical interference." *Physical Review Letters* 67, no. 3 (1991): 318–321.

Essential Reading

A Bibliography of Additional
Research Reports and Theories

Aharonov, Y., and D. Bohm. "Significance of electromagnetic potentials in the quantum theory." *Phys. Rev.* 115, no. 3 (1959).

Akimov, A. E., and G. I. Shipov. "Torsion fields and their experimental manifestations." *Journal of New Energy* 2, no. 2 (1997).

Akimov, A. E., and V. Ya. Tarasenko. "Models of polarized states of the physical vacuum and torsion fields." *Soviet Physics Journal* 35, no. 3 (1992).

Aspect, A., J. Dalibard, and F. Roger. "Experimental test of Bell's inequalities using time-varying Analyzers." *Physical Review Letters* 49 (1982): 1804–1807.

Aspect, A., and P. Grangier. "Experiments on Einstein-Podolsky-Rosen-type correlations with pairs of visible photons." In *Quantum Concepts in Space and Time*, R. Penrose and C. J. Isham, eds. Oxford: Clarendon Press, 1986.

Astin, J. A., E. Harkness, and E. Ernst. "The efficacy of 'distant healing': A systematic review of randomized trials." *Am. Journal Med.* 132 (2000).

Atmanspacher, H., H. Romer, and H. Walach. "Weak quantum theory: complementarity and entanglement in physics and beyond." *Foundations of Physics* 32 (2002).

Backster, Cleve. "Evidence of a Primary Perception in Plant Life." *Int. Journal of Parapsychology* 10, no. 4 (1968).

————. "Evidence for a Primary Perception at the Cellular Level in Plants and Animals." American Association for the Advancement of Science, Annual Meeting 2631, January 1975.

Bajpai, R. P. "Biophoton and the quantum vision of life." In *What Is Life?*, Hans-Peter Dürr, Fritz-Albert Popp, and Wolfram Schommers, eds. New Jersey, London, Singapore: World Scientific, 2002.

Barrow, John D., and Frank J. Tipler. *The Anthropic Cosmological Principle*. London and New York: Oxford University Press, 1986.

Beloussov, Lev. "The formative powers of developing organisms." In *What Is Life?*, Hans-Peter Dürr, Fritz-Albert Popp, and Wolfram Schommers, eds. New Jersey, London, Singapore: World Scientific, 2002.

Benor, Daniel J. *Healing Research,* vol. 1. London: Helix Editions, 1993.

————. "Survey of spiritual healing research." *Complementary Medical Research* 4 (1990): 9–33.

Bischof, Marco. "Introduction to integrative biophysics." In *Lecture Notes in Biophysics,* Fritz-Albert Popp and Lev V. Beloussov, eds. Dordrecht, Holland: Kluwer Academic Publishers, 2002.

Bohm, David. *Coherence and the Implicate Order.* London: Routledge & Kegan Paul, 1980.

Bohm, David, and Basil Hiley. *The Undivided Universe.* London: Routledge, 1993.

Braud, W. G. "Human interconnectedness: research indications." *Revision* 14, no. 3 (1992).

Braud, W. G., and M. Schlitz. "Psychokinetic influence on electrodermal activity." *Journal of Parapsychology* 47 (1983).

Bucher, Martin A., Alfred S. Goldhaber, and Neil Turok. "Open Universe from inflation." *Physical Review D* 52, no. 6 (15 September 1995).

Bucher, Martin A., and David N. Spergel. "Inflation in a Low-Density Universe." *Scientific American* (January 1999).

Cardeña, Etzel, Steven Jay Lynn, and Stanley Krippner. *Varieties of Anomalous Experience: Examining the Scientific Evidence.* Washington, D.C.: American Psychologcal Association, 2000.

Chaboyer, Brian, Pierre Demarque, Peter J. Kernan, and Lawrence M. Krauss. "The Age of Globular Clusters in Light of Hipparcos:

Resolving the Age Problem?" *Astrophysical Journal* 494 (10 February 1998).

Chaisson, Eric. *Cosmic Evolution: The Rise of Complexity in Nature.* Cambridge: Harvard University Press, 2000.

Clayton, Philip D. *God and Contemporary Science.* Grand Rapids, Michigan: Eerdmans, 1997.

Conforti, Michael. *Field, Form, and Fate: Patterns in Mind, Psyche and Nature.* Woodstock, Conn.: Spring Publications, 2001.

Coyle, Michael J. "Localized Reduction of the Primary Field of Consciousness as Dynamic Crystalline States." *The Noetic Journal* 3 (July 2002).

Dobzhansky, Theodosius, *Genetics and the Origin of Species,* 2nd edition. New York: Columbia University Press, 1982.

Dossey, Larry. *Recovering the Soul: A Scientific and Spiritual Search.* New York: Bantam, 1989.

———. *Healing Words: The Power of Prayer and the Practice of Medicine.* San Francisco: HarperSanFrancisco, 1993.

Duncan, A. J., and H. Kleinpoppen. "The experimental investigation of the Einstein-Podolsky-Rosen question and Bell's inequality." In *Quantum Mechanics versus Local Realism—The Einstein-Podolsky-Rosen Paradox,* F. Selleri, ed. New York: Plenum Press, 1988.

Eldredge, Niles. *Time Frames: The Rethinking of Darwinian Evolution and the Theory of Punctuated Equilibria.* New York: Simon & Schuster, 1985.

Fröhlich, H. "Long range coherence and energy storage in biological systems." *Int. Journal of Quantum Chemistry* 2 (1980).

Fröhlich, H., ed. *Biological Coherence and Response to External Stimuli.* Heidelberg: Springer Verlag, 1988.

Gazdag, László. "Superfluid mediums, vacuum spaces." *Speculations in Science and Technology* 12, no. 1 (1989).

Gilbert, S. F., J. M. Opitz, and R. A. Raff. "Resynthesizing evolutionary and developmental biology." *Developmental Biology* 173 (1996): 357–372.

Goodwin, Brian. "Development and evolution." *Journal of Theoretical Biology* 97 (1982).

———. "Organisms and minds as organic forms." *Leonardo* 22, no. 1 (1989).

———."On morphogenetic fields." *Theoria to Theory* 13 (1979).

Gould, Stephen J. "Irrelevance, submission, and partnership: the changing role of paleontology in Darwin's three centennials, and a modest proposal for macroevolution." In *Evolution from Molecules to Men,* D. Bendall, ed. Cambridge, U.K.: Cambridge University Press, 1983.

Green, Brian. *The Elegant Universe: Superstrings, Hidden Dimensions, and the Quest for the Ultimate Theory.* New York: Norton, 1999.

Gribbin, John. *In the Beginning: The Birth of the Living Universe.* New York: Little, Brown & Co., 1993.

Grinberg-Zylberbaum, J., M. Delaflor, M. E. Sanchez-Arellano, M. A. Guevara, and M. Perez. "Human communication and the electro-physiological activity of the brain." *Subtle Energies* 3, no. 3 (1993).

Grinberg-Zylberbaum, J., M. Delaflor, L. Attle, and A. Goswami, "The Einstein-Podolski-Rosen paradox." *Physics Essays* 7 (1994).

Grof, Stanislav, and Hal Zina Bennett. *The Holotropic Mind.* San Francisco: HarperSanFrancisco, 1993.

Guth, Alan H. *The Inflationary Universe: The Quest for a New Theory of Cosmic Origins.* New York: Perseus Book Group, 1997.

Hagley, E., et al. "Generation of Einstein-Podolsky-Rosen pairs of atoms." *Physical Review Letters* 79, no. 1 (1997): 1–5.

Hameroff, Stuart R. "'Funda-Mentality': Is the conscious mind subtly linked to a basic level of the universe?" *Trends in Cognitive Sciences* 2, no. 4 (1998).

Hansen, G. M., M. Schlitz, and C. Tart. "Summary of remote viewing research." In *The Mind Race,* Russell Targ and K. Harary, eds. New York: Villard, 1984.

Haroche, Serge. "Entanglement, decoherence and the quantum/classical boundary." *Physics Today* (July 1998).

Harris, W. S., M. Gowda, J. W. Kolby, C. P. Strycharz, J. L. Varck, P. G. Jones, et al. "A randomized control trial of the effects of remote, intercessory prayer on outcomes in patients admitted to the coronary care unit." *Arch. Intern. Med.* 159 (1999).

Heisenberg, Werner. *Physics and Philosophy.* New York: Harper & Row, 1985.

———. "Development of concepts in the history of quantum theory." *American Journal of Physics* 43, no. 5 (1975).

Ho, Mae-Wan, F. A. Popp, and U. Warnke, eds. *Bioelectromagnetics and Biocommunication.* Singapore: World Scientific, 1994.

Hogan, Craig J. *The Little Book of the Big Bang.* New York: Springer Verlag, 1998.

Honorton, C., R. Berger, M. Varvoglis, M. Quant, P. Derr, E. Schechter, and D. Ferrari. "Psi communication in the Ganzfeld: Experiments with an automated testing system and a comparison with a meta-analysis of earlier studies." *Journal of Parapsychology* 54 (1990).

Hoyle, Fred. *The Intelligent Universe.* London: Michael Joseph, 1983.

Josephson, B. D., and F. Pallikari-Viras. "Biological utilization of quantum nonlocality." *Foundations of Physics* 21 (1991).

Kafatos, Menas. "Non-locality, foundational principles and consciousness." *Noetic Journal* 2 (January 1999).

———. *Bell's Theorem, Quantum Theory and Conceptions of the Universe.* Dordrecht, Holland: Kluwer, 1989.

Kafatos, Menas, and R. Nadeau. *The Conscious Universe: Part and Whole in Modern Physical Theory.* New York: Springer Verlag, 1990.

Kaivarainen, Alex. "Unified Model of Bivacuum, Particles Duality, Electromagnetism, Gravitation and Time: The Superfluous Energy of Asymmetric Bivacuum." *The Journal of Non-Locality and Remote Mental Interactions* 1 (October 2002).

Krauss, Lawrence M. "The End of the Age Problem and the Case for a Cosmological Constant Revisited." *Astrophysical Journal* 501 (10 July 1998).

———. "Cosmological antigravity." *Scientific American* (January 1999).

LaViolette, Paul. *Subquantum Kinetics: A Systems Approach to Physics and Cosmology.* Alexandria, Va.: Starlane Publications, 2003.

Leslie, John. *Universes.* London and New York: Routledge, 1989.

———, ed. *Physical Cosmology and Philosophy.* New York: MacMillan, 1990.

Li, K. H. "Coherence in physics and biology." In *Recent Advances in Biophoton Research and its Applications*, F. A. Popp, K. H. Li, and Q. Gu, eds. Singapore: World Scientific Publishing, 1992.

———. "Uncertainty principle, coherence, and structures." In *On Self-Organization*, R. K. Mishra, D. Maass, and E. Zwierlein, eds. Berlin: Springer Verlag, 1994.

———. "Coherence—A Bridge between Micro- and Macro-Systems." In *Biophotonics—Non-Equilibrium and Coherent Systems in Biology, Biophysics and Biotechnology*, L. V. Belousov and F. A. Popp, eds. Moscow: Bioinform Services, 1995.

Licata, Ignazio. *Dinamica reticolare dello Spazio-Tempo* [Reticular dynamics of spacetime], Inediti. No. 27. Bologna: Andromeda, 1989.

Lieber, Michael M. "Environmentally responsive mutator systems: toward a unifying perspective." *Rivista di Biologia/Biology Forum* 91, no. 3 (1998).

———. "Force and genomic change." *Frontier Perspectives* 10, no. 1 (2001).

Lorenz, Konrad. *The Waning of Humaneness*. Boston: Little, Brown & Co., 1987.

Mallove, Eugen F. "The Self-Reproducing Universe." *Sky & Telescope* 76 (September 1988).

Maxwell, James Clerk. *A Dynamical Theory of the Electromagnetic Field*. Edited by T. F. Torrence. Edinburgh: Scottich Academic Press, 1982.

Michelson, A. A. "The relative motion of the earth and the luminiferous ether." *American Journal of Science* 22 (1881).

Mitchell, E. "Nature's mind: the quantum hologram." *International Journal of Computing Anticipatory Systems* 7 (2000).

Montecucco, Nitamo. *Cyber: La Visione Olistica*. Rome: Mediterranee, 2000.

Nelson, John E. *Healing the Split*. Albany, N.Y.: SUNY Press, 1994.

Oschman, James L. *Energy Medicine: the Scientific Basis*. London: Harcourt, 2001.

Peebles, Phillip James E. *Principles of Physical Cosmology*. Princeton, N.J.: Princeton University Press, 1993.

Penrose, Roger. *Shadows of the Mind: A Search for the Missing Science of Consciousness.* Oxford, U.K.: Oxford University Press, 2000.

Perlmuter, S., G. M. Aldering, M. Della Valle, et al. "Discovery of a Supernova Explosion at Half the Age of the Universe." *Nature* 391 (1 January 1998).

Persinger, M. A., and S. Krippner. "Dream ESP experiments and geomagnetic activity." *The Journal of the American Society for Psychical Research* 83 (1989).

Primas, Hans, H. Atmanspacher, and A. Amman, eds. *Quanta, Mind and Matter: Hans Primas in Context.* Dordrecht, Holland: Kluwer, 1999.

Puthoff, Harold. "Ground state of hydrogen as a zero-point-fluctuation-determined state." *Phys. Rev. D* 35, no. 10 (1987).

———. "Gravity as a zero-point-fluctuation force" *Phys. Rev. A* 39, no. 5 (1989).

———. "Source of vacuum electromagnetic zero-point energy." *Phys. Rev. A* 40, no. 9 (1989).

Radin, Dean. *The Conscious Universe: The Scientific Truth of Psychic Phenomena.* San Francisco: HarperEdge, 1997.

Rees, Martin. *Before the Beginning: Our Universe and Others.* New York: Addison-Wesley, 1997.

Rein, Glen. "Biological interactions with scalar energy-cellular mechanisms of action." In *Proceedings of the 7th International Association of Psychotronics Research Conference.* Atlanta, Georgia, 1988.

———. "Modulation of neurotransmitter function by quantum fields." In *Rethinking Neural Networks: Quantum Fields and Biological Data,* K. H. Pribram, ed. Hillsdale, N.J.: Erlbaum, 1993.

———. "Biological effects of quantum fields and their role in the natural healing process." *Frontier Perspectives* 7, no. 1 (1998).

Requard, Manfred. "From 'matter-energy' to 'irreducible information processing': Arguments for a paradigm shift in fundamental physics." In *Evolution of Information Processing Systems,* Kurt Hafner, ed. New York and Berlin: Springer Verlag, 1992.

Riess, Adam G., Alexei V. Filippenko, Peter Challis, et al. "Observational Evidence from Supernovae for an Accelerating Universe and a Cosmological Constant." *Astronomical Journal* 116 (September 1998).

Rothe, Gunter M. "Electromagnetic, symbiotic and informational interactions in the kingdom of organisms." In *What Is Life?*, Hans-Peter Dürr, Fritz-Albert Popp, and Wolfram Schommers, eds. New Jersey, London, Singapore: World Scientific, 2002.

Rubik, Beverly. "The Biofield Hypothesis: Its biophysical basis and role in medicine." *The Journal of Alternative and Complementary Medicine* 8, no. 6 (2002).

Schwarzschild, B. "Very distant supernovae suggest that the cosmic expansion is speeding up." *Physics Today* 51, no. 6 (1998).

Selleri, F., ed. *Quantum Mechanics versus Local Realism—The Einstein-Podolsky-Rosen Paradox*. New York: Plenum Press, 1988.

Sheldrake, Rupert. *A New Science of Life*. London: Blond & Briggs, 1981.

———. *The Presence of the Past*. New York: Times Books, 1988.

Smith, Cyril W. "Is a living system a macroscopic quantum system?" *Frontier Perspectives* 7, no. 1 (1998).

Steele, Edward J., R. A. Lindley, and R. V. Blandon. *Lamarck's Signature: New Retro-genes are Changing Darwin's Natural Selection Paradigm*. London: Allen & Unwin, 1998.

Targ, Russell, and K. Harary. *The Mind Race*. New York: Villard Books, 1984.

Taylor, R. "A gentle introduction to quantum biology." *Consciousness and Physical Reality* 1, no. 1 (1998).

Thaheld, F. H. "Proposed experiments to determine if there is a connection between biological nonlocality and consciousness." *Apeiron* 8, no. 4 (2001): 53–66.

Tiller, William A. "Subtle energies in energy medicine." *Frontier Perspectives* 4, no. 2 (1995).

Tzoref, Judah. "Vacuum kinematics: a hypothesis." *Frontier Perspectives* 7, no. 2 (1998).

———. "New aspects of vacuum kinematics." *Frontier Perspectives* 10, no. 1 (2001).

Ullman, M., and S. Krippner. *Dream Studies and Telepathy: An Experimental Approach*. New York: Parapsychology Foundation, 1970.

Wackermann, Jiri, Christian Seiter, Holger Keibel, and Harald Walach. "Correlations between brain electrical activities of two spatially separated human subjects." *Neuroscience Letters* 336 (2003).

Waddington, Conrad. "Fields and gradients." In *Major Problems in Developmental Biology,* Michael Locke, ed. New York: Academic Press, 1966.

Wagner, E. O. "Structure in the Vacuum." *Frontier Perspectives* 10, no. 2 (2001).

Weinberg, Steven. *Dreams of a Final Theory: The Search for the Fundamental Laws of Nature.* New York: Pantheon Books, 1992.

Welch, G. R. "An analogical 'field' construct in cellular biophysics: history and present status." *Progress in Biophysics and Molecular Biology* 57 (1992).

Welch, G. R., and H. A. Smith, "On the field structure of metabolic space-time." In *Molecular and Biological Physics of Living Systems,* R. K. Mishra, ed. Dordrecht, Holland: Kluwer, 1990.

Wheeler, John A. "Bits, quanta, meaning." In *Problems of Theoretical Physics,* A. Giovannini, F. Mancini, and M. Marinaro, eds. Salerno, Italy: University of Salerno Press, 1984.

———. "Quantum cosmology." In *World Science,* L. Z. Fang and R. Ruffini, eds. Singapore: World Scientific, 1987.

Whitehead, Alfred North. *An Enquiry Concerning the Principles of Natural Knowledge.* Cambridge, U.K.: Cambridge University Press, 1919.

Zeiger, Bernd F., and Marco Bischof. "The quantum vacuum and its significance in biology." Paper presented at The Third International Hombroich Symposium on Biophysics. Neuss, Germany, August 20–24, 1998 (mimeo).

Index

BOOKS OF RELATED INTEREST

Quantum Shift in the Global Brain
How the New Scientific Reality Is Changing Us and Our World
by Ervin Laszlo

Science and the Reenchantment of the Cosmos
The Rise of the Integral Vision of Reality
by Ervin Laszlo

Transcending the Speed of Light
Consciousness, Quantum Physics, and
the Fifth Dimension
by Marc Seifer, Ph.D.

A New Science of Life
The Hypothesis of Morphic Resonance
by Rupert Sheldrake

The Rebirth of Nature
The Greening of Science and God
by Rupert Sheldrake

Sacred Number and the Origins of Civilization
The Unfolding of History through the Mystery of Number
by Richard Heath

Genesis of the Cosmos
The Ancient Science of Continuous Creation
by Paul A. LaViolette, Ph.D

Stalking the Wild Pendulum
On the Mechanics of Consciousness
by Itzhak Bentov

Inner Traditions • Bear & Company
P.O. Box 388
Rochester, VT 05767
1-800-246-8648
www.InnerTraditions.com

Or contact your local bookseller